# FRP Composites for Reinforced and Prestressed Concrete Structures

# Taylor & Francis's Structural Engineering: Mechanics and Design series

Innovative structural engineering enhances the functionality, serviceability and life-cycle performance of our civil infrastructure systems. As a result, it contributes significantly to the improvement of efficiency, productivity and quality of life.

Whilst many books on structural engineering exist, they are widely variable in approach, quality and availability. With the *Structural Engineering: Mechanics and Design* series, Taylor & Francis is building up an authoritative and comprehensive library of reference books presenting the state-of-the-art in structural engineering, for industry and academia.

**Books in the series:**

**Structural Optimization: Dynamic Loading Applications**
*F.Y. Cheng and K.Z. Truman*
978-0-415-42370-0

**FRP Composites for Reinforced and Prestressed Concrete Structures**
*P. Balaguru, A. Nanni and J. Giancaspro*
978-0-415-44854-3

**Probability Theory and Statistics for Engineers**
*P. Gatti*
978-0-415-25172-3

**Modeling and Simulation-based Life-cycle Engineering**
*Edited by K. Chong, S. Saigal, S. Thynell and H. Morgan*
978-0-415-26644-4

**Innovative Shear Design**
*H. Stamenkovic*
978-0-415-25836-4

**Structural Stability in Engineering Practice**
*Edited by L. Kollar*
978-0-419-23790-7

**Earthquake Engineering**
*Y.-X. Hu, S.-C. Liu and W. Dong*
978-0-419-20590-6

**Composite Structures for Civil and Architectural Engineering**
*D.-H. Kim*
978-0-419-19170-4

# FRP Composites for Reinforced and Prestressed Concrete Structures

## A guide to fundamentals and design for repair and retrofit

## Perumalsamy Balaguru, Antonio Nanni, and James Giancaspro

CRC Press
Taylor & Francis Group
Boca Raton London New York

CRC Press is an imprint of the
Taylor & Francis Group, an **informa** business

A TAYLOR & FRANCIS BOOK

CRC Press
Taylor & Francis Group
6000 Broken Sound Parkway NW, Suite 300
Boca Raton, FL 33487-2742

First issued in paperback 2019

ISBN-13:978-0-415-44854-3(hbk)
ISBN-13:978-0-367-86573-3(pbk)

Typeset in Sabon by
Integra Software Services Pvt. Ltd, Pondicherry, India

*British Library Cataloguing in Publication Data*
A catalogue record for this book is available from the British Library

*Library of Congress Cataloging in Publication Data*
Balaguru, Perumalsamy N.
FRP composites for reinforced and prestressed concrete structures :
a guide to fundamentals and design for repair and retrofit /
Perumalsamy Balaguru, Antonio Nanni, and James Giancaspro. --
1st ed.
p. cm. -- (Spon's structural engineering mechanics and design series)
Includes bibliographical references and index.
1. Reinforced concrete construction--Maintenance and repair.
2. Prestressed concrete construction--Maintenance and repair.
3. Fiber-reinforced plastics. I. Nanni, Antonio. II. Giancaspro,
James. III. Title.
TA683.B276 2008
624.1′8340288--dc22
2008010203

Visit the Taylor & Francis Web site at
http://www.taylorandfrancis.com

and the CRC Press Web site at
http://www.crcpress.com

To our families for their encouragement, cooperation, and patience...

*Perumalsamy*
*Balaguru*

*Antonio Nanni*

*James Giancaspro*

To my family,
Balamuralee
Balasoundhari
Han Shu
Suryaprabha

To my wife,
Valeria

To my brothers
and sister

# Contents

# Preface

Fiber-reinforced polymer (FRP) is a common term used by the civil
engineering community for high-strength composites. Composites have been
used by the space and aerospace communities for over six decades and the
use of composites by the civil engineering community spans about three
decades. In the composite system, the strength and the stiffness are primarily derived from fibers, and the matrix binds the fibers together to form
structural and nonstructural components. Composites are known for their
high specific strength, high stiffness, and corrosion resistance. Repair and
retrofit are still the predominant areas where FRPs are used in the civil
engineering community. The field is relatively young and, therefore, there
is considerable ongoing research in this area. American Concrete Institute
Technical Committee 440 documents are excellent sources for the latest
information.

The primary purpose of this book is to introduce the reader to the basic
concepts of repairing and retrofitting reinforced and prestressed concrete
structural elements using FRP. Basic material properties, fabrication techniques, design concepts for strengthening in bending, shear, and confinement,
and field evaluation techniques are presented. The book is geared toward
advanced undergraduate and graduate students, professional engineers, field
engineers, and user agencies such as various departments of transportation.
A number of flowcharts and design examples are provided to facilitate easy
and thorough understanding. Since this is a very active research field, some
of the latest techniques such as near-surface mounting (NSM) techniques are
not covered in this book. Rather, the aim is to provide the fundamentals and
basic information.

The chapters in the book are arranged using the sequences followed in
textbooks on reinforced concrete. Chapter 1 provides the introduction, brief
state of history, typical field applications, and references to obtain the latest
information. In Chapter 2, the properties of common types and forms of
fiber reinforcement materials and resins are presented. Brief descriptions of
the four basic types of hybrids are also discussed. Chapter 3 deals with the

methods to prepare laminate samples, including resin mixing, hand lay-up technique, vacuum bagging, curing methods, density determination, and laminate cutting.

Chapter 4 provides the basic information on the properties of fibers and matrices used for the composite, behavior of beams and columns strengthened with the composites, design philosophies, and recent advances. The most popular uses are for (i) strengthening of reinforced and prestressed concrete beams for flexure, (ii) shear strengthening of reinforced and prestressed concrete beams, (iii) column wrapping to improve the ductility for earthquake-type loading, (iv) strengthening of unreinforced masonry walls for in-plane and out-of-plane loading, (v) strengthening for improved blast resistance, and (vi) repair of chimney or similar one-of-a-kind structures.

Chapter 5 deals with the procedures to analyze the reinforced concrete and strengthened beams and to estimate the required extra reinforcement. Even though excellent books are available for reinforced concrete design, the flexural behavior of beams is explained briefly to provide continuity in the thought process, especially in the area of tensile force-transfer from extra reinforcement to the beam.

Chapter 6 deals with the procedures to analyze prestressed concrete and strengthened beams and to estimate the required extra reinforcement. Beams with both bonded and unbonded tendons are covered. Here again, the flexural behavior of beams is explained briefly to provide continuity in the thought process, especially in the area of tensile force-transfer from extra reinforcement to the beam.

Chapter 7 presents the design procedures for shear strengthening of beams. It also presents a summary of the provisions used for reinforced and prestressed concrete, the schemes used for composite general guidelines, stress and strain limits, design procedures, and examples. As in the case of other chapters, the reader is referred to texts on reinforced concrete for detailed discussions on mechanisms and provisions.

Chapter 8 encompasses repair and retrofitting of columns. Here, the primary contribution of FRP is for confinement of concrete and the resulting improved ductility of columns. The most common retrofits are aimed at improving the earthquake resistance of columns.

Chapter 9 provides a method for efficiently and accurately assessing, in situ, the structural adequacy of reinforced and prestressed concrete building components. The guidelines allow the engineer to determine whether a specific portion of a structure has the necessary capacity to adequately resist a given loading condition. These guidelines establish a protocol for full-scale, in situ load testing including planning, executing, and evaluating a testing program, which will assist the engineer in implementing an efficient load test.

We would like to inform the readers that the tables and figures may not be exactly the same as those presented in the sources cited; modifications were made to improve clarity. Some of the illustrations have been taken from original reports, but the references cited were typically published papers.

Perumalsamy Balaguru
Antonio Nanni
James Giancaspro

# Acknowledgments

A technical book is a work of synthesis and reflection over the state-of-the-art of a specific discipline or technology. Accordingly, the authors draw from many different sources including the experiences and works of others.

It would be impossible to acknowledge all direct and indirect contributions to this very volume. In fact, over the course of many years, former and present undergraduate and graduate students have worked hard at developing the knowledge base that is the foundation of this book and many of the existing specifications and design guides published around the world. Similarly, colleagues in academia, industry, and government have also tirelessly contributed to the advancement of the state-of-the-art in the novel technology reported herein. Papers authored and coauthored by many individuals in these two groups are cited among the references and constitute a partial testimony of these contributions from others to whom the authors are deeply indebted.

Preparation of this manuscript was made possible by contributions from Ankur Jaluria, Philip Hopkins, and Paloma Alvarez.

# 1 Introduction

Fiber reinforced polymer (FRP) is a composite made of high-strength fibers and a matrix for binding these fibers to fabricate structural shapes. Common fiber types include aramid, carbon, glass, and high-strength steel; common matrices are epoxies and esters. Inorganic matrices have also been evaluated for use in fire-resistant composites. FRP systems have significant advantages over classical structural materials such as steel, including low weight, corrosion resistance, and ease of application. Originally developed for aircraft, these composites have been used successfully in a variety of structural applications such as aircraft fuselages, ship hulls, cargo containers, high-speed trains, and turbine blades (Feichtinger, 1988; Kim, 1972; Thomsen and Vinson, 2000). FRP is particularly suitable for structural repair and rehabilitation of reinforced and prestressed concrete elements. The low weight reduces both the duration and cost of construction since heavy equipment is not needed for the rehabilitation. The composites can be applied as a thin plate or layer by layer.

Even though use of FRP for civil engineering structures only started in the 1980s, a large number of projects have been carried out to demonstrate the use of this composite in the rehabilitation of reinforced and prestressed concrete structures (Hag-Elsafi *et al.*, 2001, 2004; Mufti, 2003; Täljsten, 2003). The composite has been successfully used to retrofit all basic structural components, namely, beams, columns, slabs, and walls. In addition, strengthening schemes have been carried out for unique applications such as storage tanks and chimneys. These advanced materials may be applied to the existing structures to increase any or several of the following properties:

- axial, flexural, or shear load capacities;
- ductility for improved seismic performance;
- improved durability against adverse environmental effects;
- increased fatigue life;
- stiffness for reduced deflections under service and design loads (Buyukozturk *et al.*, 2004; Täljsten and Elfgren, 2000).

*Figure 1.1* Vacuum bagging technique for rehabilitation of reinforced concrete
bridge pier.

In most cases, the FRP composites are applied manually using hand-impregnation technique. Also referred to as hand lay-up, this process involves placing (and working) successive plies of resin-impregnated reinforcement in position by hand. Squeegees and grooved rollers are used to densify the FRP structure and remove much of the entrapped air.

To avoid potential delamination failures from occurring, a denser FRP must be manufactured by removing nearly all air voids within the composite. Two methods that are capable of accomplishing this task are vacuum-assisted impregnation (vacuum bagging) and pressure bag molding (pressure bagging). Vacuum bags apply additional pressure to the composite and aid in the removal of entrapped air, as shown in Figure 1.1.

Pressure bags also invoke the use of pressure but are considerably more complex and expensive to operate. They apply additional pressure to the assembly through an electrometric pressure bag or bladder contained within a clamshell cover, which fits over a mold. However, only mild pressures can be applied with this system (May, 1987).

## 1.1 Recent advances

A number of advances have been made in the area of materials and design procedure. It is recommended that the reader seek the latest report from American Concrete Institute (ACI), Japan Concrete Institute (JCI), ISIS Canada, or CEB for the use of FRP. For example, ACI Committee 440 published a design guidelines document in October 2007 and similar documents are under preparation. JCI also updates documents frequently. ISIS publications can be obtained from the University of Manitoba. Since this is an emerging technology; changes are being made frequently to design documents to incorporate recent findings.

In the area of fibers, the major development is the reduction in the cost of carbon fibers. Other advances include development of high-modulus (up to 690 GPa) carbon fibers and high-strength glass fibers. In the case of matrix, the major advance is the development of an inorganic matrix, which is fire and UV resistant (Lyon *et al.*, 1997; Papakonstantinou *et al.*, 2001).

## 1.2 Field applications

A large number of field applications have been carried out during the last 20 years. The majority of the initial uses were in Japan, followed by applications in Europe and North America. In North America, the popular applications are in rehabilitation of bridges to improve earthquake resistance, repair and rehabilitation of parking structures, strengthening of unreinforced walls, and rehabilitation of miscellaneous structures such as tunnels, chimneys, and industrial structures such as liquid retaining tanks (ACI Committee 440, 1996). Typical examples are shown in Figures 1.2 and 1.3.

*Figure 1.2* Hand lay-up process used to strengthen reinforced concrete slab (Source: Structural Preservation Systems, 2007).

(a)

(b)

*Figure 1.3* Column wrapping with carbon tape, followed by coating of inorganic resin (a) column after cleaning; (b) column wrapped with carbon tape; (c) during final coating application; (d) column after coating (Source: Defazio *et al.*, 2006).

(c)

(d)

*Figure 1.3* (Continued).

# References

ACI Committee 440 (1996) *State-of-the-Art Report on Fiber Reinforced Plastic (FRP) Reinforcement for Concrete Structures (ACI 440 R-96)*. Farmington Hills, MI: American Concrete Institute, 68 pp.

Buyukozturk, O., Gunes, O., and Karaca, E. (2004) Progress on understanding debonding problems in reinforced concrete and steel members strengthened using FRP composites. *Construction and Building Materials*, 18(1), pp. 9–19.

Defazio, C., Arafa, M., and Balaguru, P. (2006) *Geopolymer Column Wrapping*. Report No. Mary-RU9088. New Brunswick, NJ: Center for Advanced Infrastructure and Transportation.

Feichtinger, K.A. (1988) Test methods and performance of structural core materials – I. static properties. In: *Fourth Annual ASM International/Engineering Society of Detroit Conference*, Detroit, pp. 1–11.

Hag-Elsafi, O., Alampalli, S., and Kunin, J. (2001) Application of FRP laminates for strengthening of a reinforced-concrete T-beam bridge structure. *Composite Structures*, 52(3–4), pp. 453–466.

Hag-Elsafi, O., Alampalli, S., and Kunin, J. (2004) In-service evaluation of a reinforced concrete T-beam bridge FRP strengthening system. *Composite Structures*, 64(2), pp. 179–188.

Kim, M.K. (1972) *Flexural Behavior of Structural Sandwich Panels and Design of an Air-inflated Greenhouse Structure*. PhD thesis, Rutgers, the State University of New Jersey, 105 pp.

Lyon, R.E., Balaguru, P., Foden, A.J., Sorathia, U., and Davidovits, J. (1997) Fire resistant aluminosilicate composites, *Fire and Materials*, 21(2), pp. 61–73.

May, C. (1987) Epoxy resins. In: American Society of Metals, *Engineered Materials Handbook*. Vol. 1, Ohio: American Society of Metals.

Mufti, A.A. (2003) FRPs and FOSs lead to innovation in Canadian civil engineering structures. *Construction and Building Materials*, 17(6–7), pp. 379–387.

Papakonstantinou, C.G., Balaguru, P.N., and Lyon, R.E. (2001) Comparative study of high temperature composites. *Composites Part B: Engineering*, 32(8), pp. 637–649.

Structural Preservation Systems (2007) *Strengthening of Concrete, Timber and Masonry Structures*. Hanover, MD: Structural Group.

Täljsten, B. (2003) Strengthening concrete beams for shear with CFRP sheets. *Construction and Building Materials*, 17(1), pp. 15–26.

Täljsten, B. and Elfgren, L. (2000) Strengthening concrete beams for shear using CFRP-materials: evaluation of different application methods. *Composites Part B: Engineering*, 31(2), pp. 87–96.

Thomsen, O.T. and Vinson, J.R. (2000) Design study of non-circular pressurized sandwich fuselage section using a high-order sandwich theory formulation. In: *Sandwich Construction 5, Proceedings of the 5th International Conference on Sandwich Construction* (eds H.-R. Meyer-Piening and D. Zenkert), Zürich, Switzerland.

# 2 Constituent materials

## 2.1 Introduction

Two major components of a composite are high-strength fibers and a matrix that binds these fibers to form a composite-structural component. The fibers provide strength and stiffness, and the matrix (resin) provides the transfer of stresses and strains between the fibers. To obtain full composite action, the fiber surfaces should be completely coated (wetted) with matrix. Two or more fiber types can be combined to obtain specific composite property that is not possible to obtain using a single fiber type. For example, the modulus, strength, and fatigue performance of glass-reinforced polymers (GRP) can be enhanced by adding carbon fibers. Similarly, the impact energy of carbon-fiber reinforced polymers (CFRP) can be increased by the addition of glass or aramid fibers. The optimized performance that hybrid composite materials offer has led to their widespread growth throughout the world (Hancox, 1981; Shan and Liao, 2002). In recent years, hybrid composites have found uses in a number of applications such as abrasive resistant coatings, contact lens, sensors, optically active films, membranes, and absorbents (Cornelius and Marand, 2002).

In this chapter, properties of common types and forms of fiber reinforcement materials and resins are presented.

## 2.2 Fibers

The primary role of the fiber is to resist the major portion of the load acting on the composite system. Depending on the matrix type and fiber configuration, the fiber volume fraction ranges from 30 to 75%. Strength and stiffness properties of commercially available fibers cover a large spectrum and consequently, the properties of the resulting composite have a considerable variation (Mallick, 1993). Typical fiber reinforcements used in the composite industry are glass (E- and S-glass), carbon, and aramid (Kevlar®). The properties and characteristics of these fibers as well as other fiber types such as basalt are presented in the subsequent sections.

## 2.2.1   *Glass fibers*

Glass fibers are the most common of all reinforcing fibers used in composites. Major advantages of glass fibers include low cost, high tensile strength, chemical resistance, and high temperature resistance. The disadvantages are low tensile modulus, sensitivity to abrasion while handling, relatively low fatigue resistance, and brittleness.

Glass fibers are produced by fusing silicates with silica or with potash, lime, or various metallic oxides. The molten mass is passed through microfine bushings and rapidly cooled to produce glass fiber filaments ranging in diameter from 5 to 24 μm. These filaments are then drawn together into closely packed strands or loosely packed roving. During this process, the fibers are frequently covered with a coating, known as sizing, to minimize abrasion-related degradation of the filaments (Miller, 1987; Gurit Composite Technologies, 2008).

The two most common types of glass fibers used in the fiber-reinforced plastics industry are electrical glass (also known as E-glass) and structural glass (commonly referred to as S-glass). Other less common types include chemical glass (or C-glass) and alkali-resistant glass (also known as AR-glass). Among the glass fibers, the most economical and widely used reinforcement in polymer matrix composites is E-glass. E-glass is a family of glasses with a calcium aluminoborosilicate composition and an alkali content of no more than 2.0% (Miller, 1987). Because E-glass offers good strength properties at a very low cost, it accounts for more than 90% of all glass fiber reinforcements. As its name implies, it is known for its good electrical resistance. E-glass is especially well suited for applications in which radio-signal transparency is desired, such as in aircraft radomes and antennae. It is also extensively used in computer circuit boards (Composite Basics, 2003).

S-glass has the highest tensile strength among all the glass fibers and was originally developed for missile casings and aircraft components. S-glass has a magnesium aluminosilicate composition and is more difficult to manufacture. Consequently, the cost of S-glass is considerably higher than E-glass (Miller, 1987; Gurit Composite Technologies, 2008).

Chemical glass (C-glass) has a sodalime-borosilicate composition that is utilized in corrosive environments where chemical stability is desired. It provides greater corrosion resistance to acids than E-glass. Its primary use is in surface coatings of laminates used in chemical and water pipes and tanks (Miller, 1987; Gurit Composite Technologies, 2008). Specifically developed for use in concrete, alkali-resistant glass (AR-glass) is composed of alkali-zirconium silicates. It is used in applications requiring greater chemical resistance to alkaline chemicals (bases), such as in cement substrates and concrete. Typical properties of E-, S-, C-, and AR-glass are presented in Table 2.1.

*Table 2.1* Typical properties for glass fiber types

| Glass type | Density | Tensile strength | | Modulus of elasticity | | Dielectric constant, 1 MHz at 72°F | Elongation (%) |
|---|---|---|---|---|---|---|---|
| | (g/cm³) | (MPa) | (ksi) | (GPa) | (ksi) | | |
| E-glass | 2.60 | 3445 | 500 | 72.4 | 10,500 | 6.33 | 4.8 |
| S-glass | 2.49 | 4585 | 665 | 86.9 | 12,600 | 5.34 | 5.4 |
| C-glass | 2.56 | 3310 | 480 | 68.9 | 9993 | 6.90 | 4.8 |
| AR-glass | 2.70 | 3241 | 470 | 73.1 | 10,602 | 8.10 | 4.4 |

Source: JPS, 2003; Watson and Raghupathi, 1987. Reprinted with permission of ASM International®. All rights reserved. www.asminternational.org

### 2.2.2 Carbon fibers

Carbon fibers offer the highest modulus of all reinforcing fibers. Among the advantages of carbon fibers are their exceptionally high tensile-strength-to-weight ratios as well as high tensile-modulus-to-weight ratios. In addition, carbon fibers have high fatigue strengths and a very low coefficient of linear thermal expansion and, in some cases, even negative thermal expansion. This feature provides dimensional stability, which allows the composite to achieve near zero expansion to temperatures as high as 570°F (300°C) in critical structures such as spacecraft antennae. If protected from oxidation, carbon fibers can withstand temperatures as high as 3600°F (2000°C). Above this temperature, they will thermally decompose. Carbon fibers are chemically inert and not susceptible to corrosion or oxidation at temperatures below 750°F (400°C).

Carbon fibers possess high electrical conductivity, which is quite advantageous to the aircraft designer who must be concerned with the ability of an aircraft to tolerate lightning strikes. However, this characteristic poses a severe challenge to the carbon textile manufacturer since carbon fiber debris generated during weaving may cause "shorting" or electric shocks in unprotected electrical machinery. Other key disadvantages are their low impact resistance and high cost (Amateau, 2003; Mallick, 1993).

Commercial quantities of carbon fibers are derived from three major feed-stock or precursor sources: rayon, polyacrylonitrile, and petroleum pitch. Rayon precursors, derived from cellulose materials, were one of the earliest sources used to make carbon fibers. Their primary advantage was their widespread availability. The most important drawback was the relatively high weight loss, or low conversion yield to carbon fiber, during carbonization. Carbonization is the process by which the precursor material is chemically changed into carbon fiber by the action of heat. On average, only 25% of the initial fiber mass remains after carbonization. Therefore, carbon fiber made from rayon precursors is more expensive than carbon fibers made from other materials (Hansen, 1987; Pebly, 1987).

Polyacrylonitrile (PAN) precursors constitute the basis for the majority of carbon fibers produced. They provide a carbon fiber conversion yield ranging from 50 to 55%. Carbon fiber based on PAN feedstock generally has a higher tensile strength than any other precursor. This results from a lack of surface defects, which act as stress concentrators and, consequently, reduce tensile strength (Hansen, 1987).

Pitch, a by-product of petroleum refining or coal coking, is a lower cost precursor than PAN. In addition to the relatively low cost, pitches are also known for their high carbon yields during carbonization. Their most significant disadvantage is nonuniformity from batch to batch during production (Hansen, 1987; Mallick, 1993).

Carbon fibers are commercially available with a variety of tensile moduli ranging from 30,000 ksi (207 GPa) on the low end to 150,000 ksi (1035 GPa) on the high end. With stiffer fibers, it requires fewer overall layers to achieve the optimal balance of strength and rigidity. Fiber for fiber, high-modulus and high-strength carbon weigh the same, but since high modulus is inherently more rigid, less material is required, resulting in a lighter weight composite structure for applications that require stiffer components (Competitive Cyclist, 2003).

Although use of high-modulus carbon is not very common, these fibers (>64,000 ksi or 440 GPa) have been used in a number of structures such as the London Underground subway system, one of the oldest and busiest underground railway networks in the world. High-modulus carbon fibers were successfully utilized to strengthen steel beams, cast iron struts, and girders (Moy, 2002). Table 2.2 compares some typical mechanical properties and costs of commercially available carbon fibers categorized by tensile modulus.

*Table 2.2* Typical properties of commercially available carbon fibers

| Grade | Tensile modulus | | Tensile strength | | Country of manufacture | Manufacturer | Cost per pound |
|---|---|---|---|---|---|---|---|
| | *(GPa)* | *(ksi)* | *(GPa)* | *(ksi)* | | | |
| *Standard modulus (<265 GPa) (also known as "high strength")* | | | | | | | |
| AP38-500 | 228 | 33,000 | 3.4 | 500 | Japan | Graphil | $16 |
| AP38-600 | 228 | 33,000 | 4.1 | 600 | Japan | Graphil | $24 |
| AS2 | 228 | 33,000 | 2.8 | 400 | USA | Hercules | |
| Panex 33 | 228 | 33,069 | 3.6 | 522 | USA/ Hungary | | |
| F3C | 228 | 33,069 | 3.8 | 551 | USA | | |
| T300 | 230 | 33,359 | 3.5 | 512 | USA/France/ Japan | Union Carbide/ Toray/Amoco | $26 |
| XAS | 234 | 33,939 | 3.5 | 500 | USA | Graphil/Hysol | |

| | | | | | | | |
|---|---|---|---|---|---|---|---|
| Celion | 234 | 33,939 | 3.6 | 515 | USA | Celanese/ToHo | |
| Celion ST | 234 | 33,939 | 4.3 | 629 | USA | Celanese/ToHo | |
| 34-700 | 234 | 33,939 | 4.5 | 653 | Japan/USA | | |
| TR30S | 234 | 33,939 | 4.4 | 640 | Japan | | |
| T500 | 234 | 33,939 | 3.7 | 529 | France/Japan | Union Carbide/Toray | |
| G30-500 | 234 | 34,000 | 3.8 | 550 | USA | Celion | $24 |
| G30-600 | 234 | 34,000 | 4.3 | 630 | USA | Celion | $34 |
| T700 | 235 | 34,084 | 5.3 | 769 | Japan | Toray | |
| TR50S | 235 | 34,084 | 4.8 | 701 | Japan | | |
| HTA | 238 | 34,519 | 4.0 | 573 | Germany | | |
| UTS | 240 | 34,809 | 4.8 | 696 | Japan | | |
| AS4 | 241 | 34,954 | 4.0 | 580 | USA | Hercules | $21 |
| T650-35 | 241 | 34,954 | 4.6 | 660 | USA | Amoco | $28 |
| AS5 | 244 | 35,389 | 3.5 | 508 | USA | Hercules | |
| AP38-749 | 262 | 38,000 | 5.2 | 750 | Japan | Graphil | |
| AS6 | 245 | 35,534 | 4.5 | 653 | USA | Hercules | |

*Intermediate modulus (265–320 GPa)*

| | | | | | | | |
|---|---|---|---|---|---|---|---|
| MR40 | 289 | 41,916 | 4.4 | 638 | Japan | | |
| MR50 | 289 | 41,916 | 5.1 | 740 | Japan | | |
| T1000 | 290 | 42,000 | 6.9 | 1002 | USA | Amoco | $326 |
| 42-7A | 290 | 42,000 | 5.0 | 725 | USA | Celion | $59 |
| T650-42 | 290 | 42,061 | 4.8 | 699 | USA | Amoco | $53 |
| T40 | 290 | 42,061 | 5.7 | 819 | USA | Amoco/Toray | $55 |
| T800 | 294 | 42,641 | 5.9 | 862 | France/Japan | | |
| M30S | 294 | 42,641 | 5.5 | 796 | France | | |
| IMS | 295 | 42,786 | 4.1 | 598 | Japan | | |
| G40-600 | 296 | 43,000 | 4.3 | 620 | USA | Celion | $45 |
| AP43-600 | 296 | 43,000 | 4.5 | 650 | Japan | Graphil | |
| G40-700 | 296 | 43,000 | 5.0 | 720 | USA | Celion | $47 |
| IM6 | 303 | 43,946 | 5.1 | 740 | USA | Hercules | $48 |
| IM7 | 303 | 43,946 | 5.3 | 769 | USA | Hercules | $53 |
| IM8 | 309 | 44,817 | 4.3 | 624 | USA | | |
| XIM8 | 310 | 45,000 | 5.2 | 750 | USA | Hercules | |

*High modulus (320–440 GPa)*

| | | | | | | | |
|---|---|---|---|---|---|---|---|
| XMS4 | 331 | 48,000 | 2.8 | 400 | USA | Hercules | |
| HMS4 | 338 | 49,023 | 3.1 | 450 | USA | Hercules | |
| MS40 | 340 | 49,313 | 4.8 | 696 | Japan | | |
| HMS | 341 | 49,458 | 1.5 | 220 | USA | Graphil/Hysol | |
| AP50-400 | 345 | 50,000 | 2.8 | 400 | Japan | Graphil | $55 |
| HMG50 | 345 | 50,038 | 2.1 | 300 | USA | Hitco/OCF | |
| HMA | 358 | 51,924 | 3.0 | 435 | Japan | | |
| HMU | 359 | 52,000 | 2.8 | 400 | USA | Hercules | |
| G50-300 | 359 | 52,069 | 2.5 | 360 | USA | Celion | $58 |
| AP53-650 | 365 | 53,000 | 4.5 | 650 | Japan | Graphil | $100 |

*Table 2.2* (Continued)

| Grade | Tensile modulus | | Tensile strength | | Country of manufacture | Manufacturer | Cost per pound |
|---|---|---|---|---|---|---|---|
| | *(GPa)* | *(ksi)* | *(GPa)* | *(ksi)* | | | |
| AP53-750 | 365 | 53,000 | 5.2 | 750 | Japan | Graphil | $110 |
| M40J | 377 | 54,679 | 4.4 | 640 | France/ Japan | | |
| P55 | 379 | 54,969 | 1.7 | 251 | USA | Union Carbide | |
| HR40 | 381 | 55,259 | 4.8 | 696 | Japan | | |
| M40 | 392 | 56,855 | 2.7 | 397 | Japan | | |
| PAN50 | 393 | 57,000 | 2.4 | 350 | Japan | Toray | |
| UMS2526 | 395 | 57,290 | 4.6 | 661 | Japan | | |
| *Ultra high modulus (~440 GPa)* | | | | | | | |
| UMS3536 | 435 | 63,091 | 4.5 | 653 | Japan | | |
| M46J | 436 | 63,236 | 4.2 | 611 | Japan | | |
| HS40 | 441 | 63,962 | 4.4 | 638 | Japan | | |
| UHMS | 441 | 63,962 | 3.5 | 500 | USA | Hercules | $325 |
| GY70 | 483 | 70,053 | 1.5 | 220 | USA | Celion/ Celanese | $750 |
| P75 | 517 | 74,985 | 2.1 | 300 | USA | Union Carbide | |
| Thornel 75 | 517 | 74,985 | 2.5 | 365 | USA | Union Carbide | |
| GY80 | 572 | 83,000 | 5.9 | 850 | USA | Celion | $850 |
| P100 | 724 | 105,007 | 2.2 | 325 | USA | Union Carbide | |

Source: Amateau, 2003; Hansen, 1987; Gurit Composite Technologies, 2008.

### 2.2.3 *Aramid fibers*

Aramid fiber is a synthetic organic polymer fiber (an aromatic polyamide) produced by spinning a solid fiber from a liquid chemical blend. Aramid fiber is bright golden yellow and is commonly known as Kevlar®, its DuPont trade name. These fibers have the lowest specific gravity and the highest tensile strength-to-weight ratio among the reinforcing fibers used today. They are 43% lighter than glass and approximately 20% lighter than most carbon fibers. In addition to high strength, the fibers also offer good resistance to abrasion and impact, as well as chemical and thermal degradation. Major drawbacks of these fibers include low compressive strength, degradation when exposed to ultraviolet light and considerable difficulty in machining and cutting (Mallick, 1993; Smith, 1996; Gurit Composite Technologies, 2008).

Kevlar® was commercially introduced in 1972 and is currently available in three different types:

- Kevlar®49 has high tensile strength and modulus and is intended for use as reinforcement in composites.

- Kevlar®29 has about the same tensile strength, but only about two-thirds the modulus of Kevlar®49. This type is primarily used in a variety of industrial applications.
- Kevlar® has tensile properties similar to those of Kevlar®29 but was initially designed for rubber reinforcement applications.

Table 2.3 shows that Kevlar®29 is nearly identical to Kevlar®49, with the exception of tensile modulus. The specific modulus in Table 2.3 is simply the modulus of the material divided by the material density, and is a measure of the stiffness of a material per unit weight. Materials with high specific moduli provide the lowest deflection for the lowest weight. Along with its tendency to yield in compression, Kevlar® exhibits a higher elongation at failure than glass and carbon fibers, resulting in a tougher and less brittle fiber than other commonly used reinforcing fibers. However, this high toughness is responsible for the significant difficulties in cutting and machining operations. Specially developed ultrasonic tools are needed to cut materials containing or composed of aramid fibers (Schwartz, 1985).

Kevlar® has been extremely successful in a variety of applications including premium tire cords, marine cordage, military body armor, oxygen bottles, high-pressure rocket casings, propeller blades, and in engine cowlings and wheel pants of aircraft, which are subject to damage from flying gravel (Smith, 1996).

*Table 2.3* Comparative fiber mechanical properties

| Property | Basalt | High-strength carbon | Kevlar® 29 | Kevlar® 49 | E-glass | S-glass |
|---|---|---|---|---|---|---|
| Fiber density | | | | | | |
| (lb/in.$^3$) | 0.098 | 0.063 | 0.052 | 0.052 | 0.092 | 0.090 |
| (g/cm$^3$) | 2.7 | 1.75 | 1.44 | 1.44 | 2.55 | 2.49 |
| Break elongation | | | | | | |
| (%) | 3.10 | 1.25 | 4.40 | 2.90 | 4.70 | 5.60 |
| Tensile strength | | | | | | |
| (ksi) | 702 | 450 | 525 | 525 | 500 | 683 |
| (GPa) | 4.84 | 3.1 | 3.62 | 3.62 | 3.45 | 4.71 |
| Specific tensile strength[a] | | | | | | |
| (10$^6$ in.) | 7.2 | 7.1 | 10.1 | 10.1 | 5.4 | 7.6 |
| (10$^7$ cm) | 1.8 | 1.8 | 2.5 | 2.5 | 1.4 | 1.9 |
| Tensile modulus | | | | | | |
| (ksi × 10$^3$) | 12.9 | 32 | 12 | 18 | 10 | 12.4 |
| (GPa) | 89 | 221 | 83 | 124 | 69 | 85 |
| Specific tensile modulus[a] | | | | | | |
| (10$^8$ in.) | 1.3 | 5.1 | 2.3 | 3.5 | 1.1 | 1.4 |
| (10$^8$ cm) | 3.3 | 12.6 | 5.7 | 9.0 | 2.7 | 3.5 |

Source: Albarrie, 2003; Schwartz, 1985.

Note: [a] Specific property = property/material density.

## 2.2.4  *Basalt fibers*

Basalt fiber is a unique product derived from volcanic material deposits. Basalt is an inert rock, found in abundant quantities, and has excellent strength, durability, and thermal properties. The density of basalt rock is between 175 and 181 lb/ft³ (2800 and 2900 kg/m³). It is also extremely hard $-8$ to 9 on the Mohs Hardness Scale (diamond $= 10$). Consequently, basalt has superior abrasion resistance and is often used as a paving and building material.

While the commercial applications of cast basalt have been well known for a long time, it is less known that basalt can be formed into continuous fibers possessing unique chemical and mechanical properties. The fibers are manufactured from basalt rock in a single-melt process and are better than glass fibers in terms of thermal stability, heat and sound insulation properties, vibration resistance, as well as durability. Basalt fibers offer an excellent economic alternative to other high-temperature-resistant fibers and are typically utilized in heat shields, composite reinforcements, and thermal and acoustic barriers (Albarrie, 2003). Table 2.3 compares some typical mechanical properties of basalt fibers with Kevlar®, high-strength carbon, E-, and S-glass. It can easily be seen that the basalt fibers have the highest tensile strength compared to the other fibers.

## 2.2.5  *Comparison of fiber properties*

Figure 2.1 presents a simple cost comparison for the most common types of fiber reinforcements. The prices are based upon continuous tows (rovings) of each fiber type. It shows that E-glass is the most economical type of fiber available today. In addition, the figure illustrates that higher modulus carbon fibers are the most expensive. Figure 2.2 compares the tensile modulus (stiffness) of typical fibers with that of traditional metals used in engineering applications. The bar chart shows that ultra high-modulus (UHM) carbon fiber has a modulus three times that of steel and standard modulus carbon fiber has a modulus twice that of aluminum.

Consider the tensile strength of common fiber reinforcements when compared to that of titanium, steel, and aluminum (Figure 2.3). The tensile strength of the fibers considered here far exceed that of aluminum by as much as 400%. For the most part, carbon, Kevlar®, and fiberglass also exceed the strength of steel by as much as two times. The specific strength of all of the fibers surpasses that of the metals by as much as ten times. Carbon, Kevlar®, and fiberglass fibers offer superior strength at a lower weight when compared to metals.

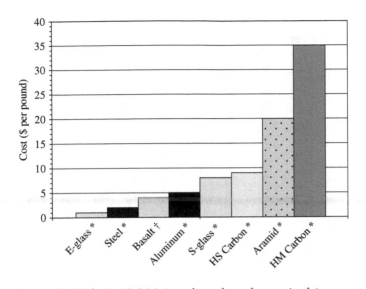

*Figure 2.1* Relative ROM (rough order of magnitude) raw material costs (Source: Advanced Composites Inc., 2003; Plant of Insulation Materials, Ltd, 2003).

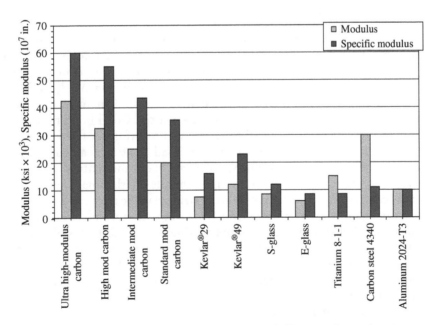

*Figure 2.2* Tensile modulus (stiffness) of typical fibers and metals (Source: Composite Tek, 2003).

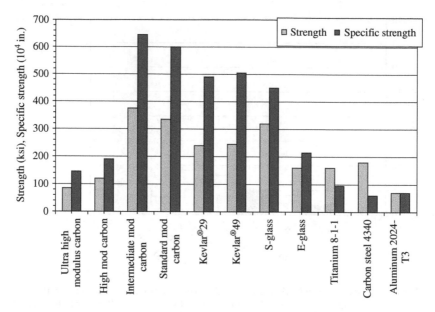

*Figure 2.3* Tensile strength of typical fibers and metals (Source: Composite Tek, 2003).

### 2.2.6 *Fiber sizing*

Surface sizings, also known as finishing agents, coupling agents, or size coatings, are an essential factor in fibrous composite technologies. The term "sizing" refers to any surface coating applied to a fiber reinforcement to protect it from damage during processing. The sizing agent also improves the fiber surface wetability with the matrix (resin). This in turn results in a stronger bond between the fiber and the matrix. Good bond between the fiber and the matrix is essential for effective stress transfer from the fiber to the matrix and vice versa. The interfacial bond created by a coupling agent allows a better shear stress transfer between fibers and matrix, which, consequently, improves the tensile strength and the interlaminar shear strength of the composite. Table 2.4 presents the sizing classifications and their typical functions. Many commercially available sizings are formulated to be multifunctional. For instance, a glass fiber sizing may consist of a film-forming polymer to produce a uniform protective coating as well as an organofunctional silane to facilitate adhesion. The degree of bond improvement (or lack thereof) is critically dependent upon the compatibility of the sizing with the matrix (Bascom, 1987; Mallick, 1993).

Although the main function of the sizing is to improve the mechanical properties of fibrous composites, they can present a number of significant disadvantages. Some experts have considered sizing as a "necessary evil" in that they are needed in one stage of processing but can hinder and interfere

*Table 2.4* Sizing classifications and functions

| Sizing type | Purpose | Example | Comments |
| --- | --- | --- | --- |
| Film-forming organics and polymers | To protect the reinforcement during processing | Polyvinyl alcohol (PVA), polyvinyl acetate (PVAc) | The polymer is formulated to wet-spread to form a uniform coating that is applied to aid processing but later may be removed by washing or heat cleaning, for example, fugitive sizing |
| Adhesion promoters | To improve composite mechanical properties and/or moisture resistance | Silane coupling agents (vinyl, glycidyl, and methacryl silane) | Principally used on inorganic reinforcement, for example, glass fiber |
| Interlayer | To enhance composite properties by creating an interphase between matrix and reinforcement | Elastomeric coating | Not in commercial use yet |
| Chemical modifiers | React to form protective coating | Silicon carbide on boron fibers | |

Source: Bascom, 1987. Reprinted with permission of ASM International®. All rights reserved. www.asminternational.org

with subsequent processing. For instance, when carbon fiber is manufactured, sizing must be applied to the fiber tow to prevent the individual filaments from damage when contacting one another or with eyelets or guides during the weaving process. However, this same sizing may actually bond the filaments together, preventing uniform impregnation of the tows by the resin. In addition, sizing treatments may interfere with or adversely affect composite mechanical properties or durability. For example, a sizing that holds the filaments in a bundle so that the strand (tow) can be chopped for discontinuous fiber composites will interfere with later efforts to disperse the fibers during extrusion or injection molding (Bascom, 1987).

### 2.2.7 Forms of reinforcement

All the fiber types are available in a variety of forms to serve a wide range of processes and end-product requirements. Fibers supplied as reinforcement include continuous spools of tow (carbon), roving (glass), milled fiber,

*Table 2.5* Various forms of fiber reinforcements

| Reinforcement form | Specification | Common applications |
|---|---|---|
| Chopped strand | Strands chopped 6 to 50 mm (1/4 to 2″) long | Injection molding; matched die; can improve fracture resistance and lessen brittleness |
| Chopped strand mat | Non-woven random mat consisting of chopped strands | Hand lay-up; resin transfer molding (RTM); very economical |
| Continuous strand mat | Non-woven random mat consisting of continuous strands | Resin transfer molding (RTM); cold press molding; random nature of strands provide excellent random-direction mechanical properties |
| Filament | Fibers as initially drawn; smallest unit of material | Processed further before use |
| Milled fiber | Continuous strands hammer-milled into short lengths 0.8 to 3 mm (1/32 to 1/8″) in length; powder-like consistency | Compounding; casting; reinforced reaction injection molding (RRIM) |
| Non-woven fabric | Felt type mat made from entangling bonded filaments | Hand lay-up; spray-up; RTM |
| Roving | Long and narrow strands bundled together like rope, but not twisted | Filament winding; spray-up; pultrusion |
| Spun roving | Continuous single strand looped and twisted | Processed further before use |
| Surfacing mat | Random mat of monofilaments | Hand lay-up; matched die; pultrusion; used to create a smooth surface on a laminate |
| Tow | Untwisted filaments gathered together into one continuous bundle | Hand lay-up; processed further before use |
| Woven fabric | Planar cloth woven from 2 or more sets of yarns, fibers, rovings, or filaments | Hand lay-up; prepreg |
| Woven roving | Strands woven like fabric but coarser and heavier | Hand lay-up; spray-up; RTM; commonly sandwiched between layers of mat |
| Yarn | Continuous length of interlocked fibers or twisted strands (often treated with a chemical sizing) | Processed further before use |

Source: Greene 1999.

chopped strands, chopped or thermo-formable mat, and woven fabrics. Reinforcement materials can be tailored with unique fiber architectures and be preformed (shaped) depending on the product requirements and manufacturing process. Table 2.5 provides a simple summary of the various forms of fiber reinforcements. These forms are discussed in the following sections.

### 2.2.7.1 Filament

A filament is an individual fiber as drawn during processing (drawing and spinning). It can be considered as the smallest unit of fiber reinforcement. Depending on the material, the filament diameter can range from 1 to 25 μm. Table 2.6 presents the standard filament diameter nomenclature as used in the fiberglass industry. It is standard practice to use a specific alphabet designation when referring to a specific filament diameter. Very fine fibers typically used in textile applications, range from AA to G. Conventional composite reinforcements consist of filaments with diameters ranging from G to U. Individual filaments are rarely used as reinforcement; they are typically gathered into strands of fibers (either continuous or chopped) for use in fibrous composites (Watson and Raghupathi, 1987).

### 2.2.7.2 Yarn

A yarn is a generic term for a closely associated bundle of twisted filaments, continuous strand of fibers, or strands in a form suitable for knitting,

*Table 2.6* Standard filament diameter nomenclature

| Alphabet | Filament Diameter | | Alphabet | Filament Diameter | |
|---|---|---|---|---|---|
| | ($\mu m$) | ($10^{-4}$ in.) | | ($\mu m$) | ($10^{-4}$ in.) |
| AA | 0.8 – 1.2 | 0.3 – 0.5 | K | 12.7 – 14.0 | 5.0 – 5.5 |
| A | 1.2 – 2.5 | 0.5 – 1.0 | L | 14.0 – 15.2 | 5.5 – 6.0 |
| B | 2.5 – 3.8 | 1.0 – 1.5 | M | 15.2 – 16.5 | 6.0 – 6.5 |
| C | 3.8 – 5.0 | 1.5 – 2.0 | N | 16.5 – 17.8 | 6.5 – 7.0 |
| D | 5.0 – 6.4 | 2.0 – 2.5 | P | 17.8 – 19.0 | 7.0 – 7.5 |
| E | 6.4 – 7.6 | 2.5 – 3.0 | Q | 19.0 – 20.3 | 7.5 – 8.0 |
| F | 7.6 – 9.0 | 3.0 – 3.5 | R | 20.3 – 21.6 | 8.0 – 8.5 |
| G | 9.0 – 10.2 | 3.5 – 4.0 | S | 21.6 – 22.9 | 8.5 – 9.0 |
| H | 10.2 – 11.4 | 4.0 – 4.5 | T | 22.9 – 24.1 | 9.0 – 9.5 |
| J | 11.4 – 12.7 | 4.5 – 5.0 | U | 24.1 – 25.4 | 9.5 – 10.0 |

Source: Watson and Raghupathi, 1987. Reprinted with permission of ASM International®. All rights reserved. www.asminternational.org

weaving, or otherwise intertwining to form a textile fabric. Yarn occurs in the following forms:

- Spun yarn is a number of fibers twisted together.
- Zero-twist yarn is a number of filaments laid together without twist.
- Twisted yarn is a number of filaments laid together with a degree of twist.
- Monofilament is a single filament with or without twist.
- The last form is simply a narrow strip of material, such as paper, plastic film, or metal foil, with or without twist, intended for use in a textile construction (Celanese Acetate LLC, 2001).

Yarns have varying weights described by their "Tex" (the weight in grams of 1000 linear meters) or "denier" (the weight in pounds of 10,000 yds or the weight in grams of 9000 m), the lower the denier, the finer the yarn. The typical Tex range is usually between 5 and 400. Most yarns, especially glass, follow an internationally recognized terminology as shown in the example of Table 2.7 (Pebly, 1987; Gurit Composite Technologies, 2008).

### 2.2.7.3  Tow

A tow is an untwisted bundle of continuous filaments. Also known as a continuous strand, or an "end," it is commonly used when referring to manufactured fibers, especially carbon. Tow designations are based upon the number of thousands of fibers. For example, a "12k HMC Tow" refers to a high modulus carbon tow consisting of 12,000 fibers. Tows are sold by weight (pounds or kilograms) and are typically wound onto a spool, as shown in Figure 2.4 (Pebly, 1987).

### 2.2.7.4  Roving

Unlike yarns, a roving is a *loosely* assembled bundle of untwisted parallel filaments or strands. Each filament diameter in a roving is the same, and is usually 13–24 μm. Rovings have varying weights, and the Tex range

*Table 2.7* Example of terminology used to identify glass yarn

| Glass type | Yarn type | Filament diamter (μm) | Strand weight (Tex) | Single strand twist | No. of strands | Multi-strand twist | No. turns per meter |
|---|---|---|---|---|---|---|---|
| E | C | 9 | 34 | Z | X2 | S | 150 |

Source: Gurit Composite Technologies, 2008.

Notes: E = electrical; C = continuous; S = high strength; Z = clockwise; S = anticlockwise.

*Figure 2.4* Carbon tow wound around spools (Source: Zoltek, 2008).

is usually between 300 and 4800. If filaments are gathered together directly after the melting process, the resultant fiber bundle is known as a direct roving. If several strands are assembled together after the glass is manufactured, they are known as an assembled roving. Assembled rovings usually have smaller filament diameters than direct rovings, providing better wet-out and mechanical properties, but they can suffer from catenary problems (unequal strand tension), and are usually more expensive due to more complex manufacturing processes (Gurit Composite Technologies, 2008).

Rovings are typically used in continuous molding operations, such as filament winding and pultrusion. In addition, rovings can be preimpregnated with a thin layer of resin to form prepregs (ready-to-mold material that can be stored until time of use). When designating reinforcement weights of rovings, the unit of measure is "yield," which is defined as the number of linear yards of roving per pound. Thus, "162 yield roving" equals 162 yds per pound (Celanese Acetate LLC, 2001). Figure 2.5 shows glass roving spun onto large spools.

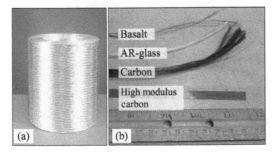

*Figure 2.5* (a) Glass roving (Source: Saint Gobain, 2003); (b) Comparison of fiber tows and roving.

## 2.2.7.5  *Chopped strands*

Chopped strands are produced by cutting continuous strands into short lengths. The ability of the individual filaments to remain together during or after the cutting process depends on the type and amount of sizing applied during manufacturing. Strands of high integrity that remain together are referred to as being "hard" while those that separate more easily are called "soft." Chopped strands, 0.12–0.47 in. (3–12 mm) in length, are typically used in injection molding processes. Chopped strand mats are usually made with longer strands, which measure up to 2 in. (50 mm). Chopped strands of carbon, glass, aramid, and basalt are commercially available and are sold by weight (Mallick, 1993). Pictures of chopped AR-glass strands are shown in Figure 2.6.

## 2.2.7.6  *Milled fibers*

Milled fibers are produced by grinding continuous strands in a hammer mill into very short lengths. Fiber lengths typically range from particulates to screen opening dimensions ranging from 0.04 to 0.12 in. (1 to 3 mm). They are primarily used in the plastics industry as inexpensive filler. Although they provide increased stiffness and dimensional stability to plastics, they do not provide significant reinforcement value. Typical applications include reinforced reaction injection molding (RRIM), phenolics, and potting compounds (Mallick, 1993; Watson and Raghupathi, 1987).

*Figure 2.6* Chopped AR-glass strands.

### 2.2.7.7 Fiber mats

A fiber mat, also known as "omni-directional reinforcement" is randomly oriented fibers held together with a small amount of adhesive binder. Fiber mats can be used for hand lay-up as prefabricated mat or for the spray-up process as chopped strand mat. The essential points of fiber mats are:

- cost much less than woven fabrics and are about 50% as strong;
- require more resin to fill interstices and more vacuum to remove air;
- used for inner layers and help in filling complex fabrics;
- high permeability and easy to handle;
- low stiffness and strength and no orientation control;
- mechanical properties are less than other reinforcements;
- used in non-critical applications.

Three typical types mat reinforcements are:

1 randomly oriented chopped filaments (chopped strand mat)
2 swirled filaments loosely held together with binder (continuous strand mat)
3 very thin mats of highly filamentized glass (surfacing mat).

A chopped strand mat is a non-woven material composed of chopped fiberglass of various lengths randomly dispersed to provide equal distribution in all directions and held together by a resin-soluble binder. Chopped strand mats are commonly used in laminates due to ease of wet-out, good bond provided between layers of woven roving or cloth, and comparatively low cost. Chopped strand mat is categorized by weight per square foot and is sold by the running meter or in bulk by weight in full rolls (Gurit Composite Technologies, 2008; Watson and Raghupathi, 1987).

A continuous strand mat is similar to a chopped strand mat, except that the fiber is continuous and is made by swirling strands of continuous fiber onto a belt, spraying a binder over them, and then drying the binder. Both hand lay-up and spray-up produce plies with equal physical properties and good interlaminar shear strength. This is a very economical way to build up thickness, especially with complex molds. Continuous strand mats are usually designated in gram per square meter (Gurit Composite Technologies, 2008; Watson and Raghupathi, 1987).

A surfacing mat is a very fine mat made from glass or carbon fiber and is used as a top layer in a composite to provide a more aesthetic surface by hiding the glass fibers of a regular woven fabric. A surfacing mat is similar in appearance to the chopped strand mat but is much finer (usually 180 to 510 μm). It is composed of fine fiberglass strands of various lengths randomly dispersed in all directions and held together by a resin-soluble binder. It is characterized by uniform fiber dispersion, a smooth and soft surface, low binder content, fast resin impregnation, and conforming well to molds. This material is used to provide a resin-rich layer in liquid or chemical holding tanks, or as a reinforcement for layers of gelcoat (a quick-setting resin applied to the surface of a mold and gelled before lay-up) (Gurit Composite Technologies, 2008; Watson and Raghupathi, 1987).

### 2.2.8   Fabrics

A fabric is defined as a manufactured assembly of long fibers of carbon, aramid, glass, or other fibers, or a combination of these, to produce a flat sheet of one or more layers of fibers. These layers are held together either by mechanical interlocking of the fibers themselves or with a secondary material to bind these fibers together and hold them in place, giving the assembly sufficient integrity for handling. Consequently, fabrics are the preferred choice of reinforcement since the fibers are in a more convenient format for the design engineer and fabricator. Fabric types are categorized by the orientation of the fibers used, and by the various construction methods used to hold the fibers together (Cumming, 1987; Gurit Composite Technologies, 2008). Before each type of fabric architecture is discussed, some relevant terminology is presented.

### 2.2.8.1   Terminology

The weight of a dry fabric is usually represented by its area density, or weight per unit area (usually just called "weight"). The most common unit of measure is ounces per square yard, often simply abbreviated as "ounces." Thus, a fabric with a weight of "5.4 oz" really has an area density of $5.4 \, oz/yd^2$.

Each fabric has its own pattern, often called the construction, and is an $x$, $y$ coordinate system (Figure 2.7). Some of the yarns run in the direction of the roll ($y$-axis or 0°) and are continuous for the entire length of the roll. These are the warp yarns and are usually called *ends*. The $y$-axis is the long axis of the roll and is typically 33 to 164 yards (30–150 m). The short yarns, which run crosswise to the roll direction ($x$-axis or 90°), are called the *fill* or *weft* yarns (also known as picks). Therefore, the $x$-direction is the roll width and is usually 39 to 118 in. (1 to 3 m).

Fabric *count* refers to the number of warp yarns (ends) and fill yarns (picks) per inch. For example, a "24 × 22 fabric" has 24 ends in every inch of fill direction and 22 picks in every inch of warp direction. It is important to note that warp yarns are counted in the fill direction, while fill yarns are counted in the warp direction. Two other important terms are drape and bias. Drape refers to the ability of a fabric to conform or fit into a contoured surface, and bias represents the angle of the warp and weft threads, usually 90° but can be 45° (Cumming, 1987; Gurit Composite Technologies, 2008).

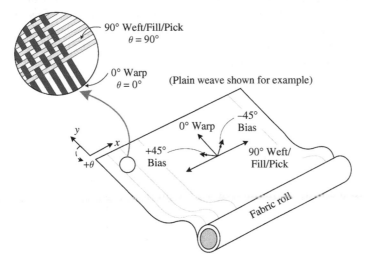

*Figure 2.7* Fabric orientation (Source: Smith, 1996).

### 2.2.8.2 Unidirectional fabrics

A fabric made with a weave pattern designed for strength in only one direction is termed unidirectional. The pick count of a unidirectional fabric is very small and most of the yarns run in the warp direction. Pure unidirectional construction implies no structural reinforcement in the fill direction, although enough warp fibers are included in the weave to ensure ease of handling (Figure 2.8). The small amount of fiberglass fill in the two fabrics of Figure 2.8 can be seen traveling horizontally in the *x*-direction. Ultrahigh-strength/modulus material, such as carbon fiber, is sometimes used in this form for specific application. Material widths are generally limited due to the

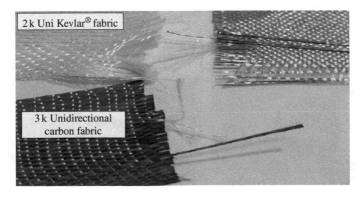

*Figure 2.8* Unidirectional Kevlar® and unidirectional carbon fabrics.

difficulty of handling and wet-out. Consequently, unidirectional fabrics are commonly manufactured in tape form or narrow rolls (less than a few inches wide). Typical applications of unidirectionals include highly loaded designed composites, such as aircraft components or race boats. Entire hulls will be fabricated from unidirectional reinforcements if an ultra high performance laminate is desired (Cumming, 1987; Smith, 1996; Gurit Composite Technologies, 2008).

### 2.2.8.3   Weave types

The weave describes how the warp and fill yarns in a fabric are interlaced. Weave determines the ability to drape and isotropy of strength (some weaves are biased to the warp or fill direction). The most popular weaves are plain, twill, basket weave, harness satin, and crowfoot satin (Figure 2.9).

The plain weave (0/90°) is the most common weave construction used in the composites industry. As shown in Figures 2.9 and 2.10, construction of

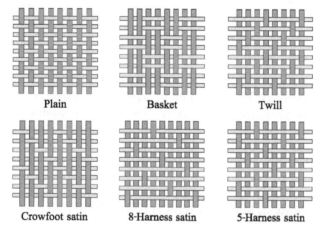

| Plain | Basket | Twill |
| Crowfoot satin | 8-Harness satin | 5-Harness satin |

*Figure 2.9* Basic weave types (Source: Dominguez, 1987).

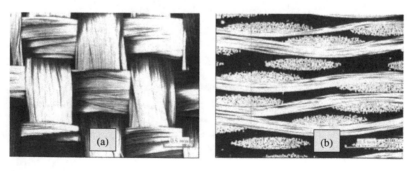

*Figure 2.10* Plain weave: (a) plan view, (b) side view (Source: Pilling, 2008).

the plain weave essentially requires only four weaving yarns: two warps and two fills. This basic unit is termed as the *pattern repeat*. The interlacing yarns follow the same simple pattern: one over and one under. For a simple plain-woven fabric, half of the fibers are in the warp (0°) orientation and the other half is in the fill (90°) direction. This weave type is highly interlaced and is, consequently, one of the tighter fabric constructions and very resistant to in-plane shear movement. The fabric is symmetrical, with good stability and reasonable porosity.

The strength of plain weaves is somewhat compromised due to the severe "pre-buckling" already present in the fabric. Fibers provide their greatest strength when they are perfectly straight. The frequent over-and-under cross-ing of the threads in a plain weave induces a slight curvature in the fibers, essentially "pre-buckling" the tows even before any load has been applied. This fiber undulation reduces the strength of plain weave types, though they are still adequate for many applications. Plain weaves are the most difficult of the weaves to drape, and the high level of fiber crimp results in relatively low mechanical properties compared with the other weave styles. With large fibers (high Tex), the plain weave gives excessive crimp and therefore is not used for very heavy fabrics. Plain weaves are typically used for flat lamin-ates, printed circuit boards, narrow fabrics, molds, and covering wood boats (Cumming, 1987; Gurit Composite Technologies, 2008).

In a twill weave, one or more warp fibers alternately weave over and under two or more weft fibers in a regular repeated manner. This produces the visual effect of a straight or broken diagonal "rib" to the fabric (Figure 2.11(a)). Twills are characterized by the diagonal pattern that is formed by the weave. This optical illusion often confuses fabricators into laying-up the material 45° off the desired fiber orientation. The twill is formed when the weft passes over warps 1 and 2 and under warps 3 and 4, and in the next pass, the shuttle of the loom passes over warps 2 and 3 and under warps 4 and 5. The advantage of the twill is the fewer number of times that the fibers go under and over one another. Better wet-out and drape is possible with twill weave compared with the plain weave with only a small reduction in stability. With reduced crimp, the fabric also has a smoother surface and slightly better mechanical

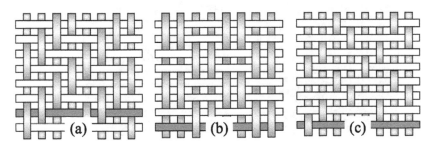

*Figure 2.11* (a) Twill weave, (b) basket weave, (c) satin weave (Source: Gurit Composite Technologies, 2008).

properties. Twills are known for being the most drapable weave and are often used for complex shapes in both vacuum bagging and wet lay-up applications. Carbon fiber twill material is often chosen for its aesthetic appearance.

The basket weave is fundamentally similar to plain weave but two yarns are grouped together and woven in an over-two-under-two fashion (Figure 2.11(b)). An arrangement of two warps crossing two wefts is designated as a "2 × 2 basket," but the arrangement of fiber need not be symmetrical. The weave can be varied where four yarns are woven over-four-under-four, hence, a "4 × 4 basket weave." Other variations such as 8 × 2 and 5 × 4 are possible. A basket weave is flatter and, through less crimped, stronger and more pliable than a plain weave but less stable. Basket weaves have less prebuckling because the yarns do not alternate over-and-under as often. It can be used on heavyweight fabrics made with thick (high Tex) fibers without excessive crimping.

The satin weaves represent a family of constructions with a minimum of interlacing. In these fabrics, the weft yarns periodically skip, or "float," over several warp yarns as shown in Figure 2.11(c). Satin weaves are fundamentally twill weaves modified to produce fewer intersections of warp and weft. The satin weave repeat is $n$ yarns long and the float length is $(n - 1)$ yarns. Therefore, there is only one interlacing point per pattern repeat per yarn. The floating yarns that are not being woven into the fabric create considerable looseness or suppleness. As a result, the satin weave construction has low resistance to shear distortion but is easily molded (draped) over compound curves. This is one of the important reasons why engineers frequently use satin weaves in aerospace applications.

The "harness" number used in the designation (typically 4, 5, or 8) is the total number of fibers crossed and passed under, before the fiber repeats the pattern. As the number of harnesses increases, so do the float lengths and the degree of looseness, increasing the difficulty during handling operations. For example, consider the 5-harness satin weave shown in Figure 2.12. In this weave, each yarn goes over 4 and under 1 yarn in both directions. Figure 2.13

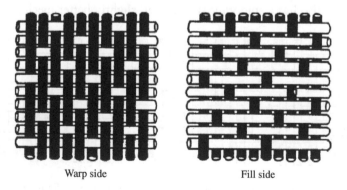

Warp side                    Fill side

*Figure 2.12* Five-harness satin weave construction (Source: DOD, 1996).

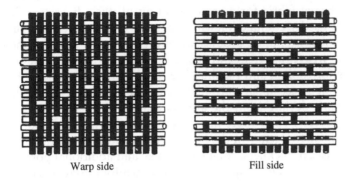

Warp side                          Fill side

*Figure 2.13* Eight-harness satin weave construction (Source: DOD, 1996).

shows the 8-harness satin weave in which each yarn goes over 7 and under 1 yarn in both directions. A crowfoot weave is a form of satin weave with a different stagger in the repeat pattern (Figure 2.11).

Satin weaves are known to be very flat, have good wet-out properties, low crimp, and a high degree of drape. Satin weaves allow fibers to be woven in the closest proximity and can produce fabrics with a close, 'tight' weave. The low stability and asymmetry of satin weaves need to be considered. The asymmetry stems from one face of the fabric having fiber running predominantly in the warp direction while the other face has fibers running predominantly in the weft direction. Figures 2.12 and 2.13 illustrate this asymmetry in the 5- and 8-harness satin weaves, respectively. When assembling multiple layers of these fabrics, care must be taken to ensure that stresses are not built into the component through this asymmetric effect.

Two less common weaves are the leno and mock leno. The leno (also known as locking leno) is a form of plain weave in which two adjacent warp fibers are twisted around each fill yarn to form a spiral pair, essentially "locking" each fill yarn in place (Figure 2.14(a)). The advantages of the leno

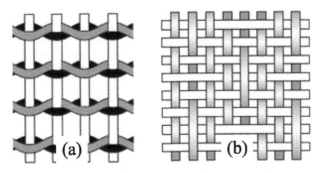

*Figure 2.14* (a) Leno, (b) Mock leno weaves (Source: Gurit Composite Technologies, 2008).

weave are its high stability in "open" fabrics, which have a low fiber count, and it provides heavy fabric for a rapid build-up of plies. Fabrics in leno weave are usually used in conjunction with other weave styles because, if used alone, their openness would not produce an effective composite component. Fabrics with leno weaves are frequently used as an inner core for the support of thin coatings for tooling and repairs. The mock leno is a version of the plain weave in which occasional warp fibers, at regular intervals but usually several fibers apart, deviate from the alternate under-over interlacing and instead interlace every two or more fibers. As shown in Figure 2.14(b), this happens with similar frequency in the fill direction, and the overall effect is a fabric with increased thickness, rougher surface, and additional porosity.

Table 2.8 presents a simple comparison of properties for the weaves presented in this section. It can be seen that each fabric has its advantages and disadvantages. For example, consider the satin and leno weaves. While the satin weave has excellent drape and poor stability, the leno weave has excellent stability and very poor drape. The fabric must be inherently stable enough to be handled, cut, and transported to the mold, yet pliable enough to conform to the mold shape and contours. If properly designed, the fabric will allow for quick wet-out and will stay in place once the resin is applied (Cumming, 1987; Gurit Composite Technologies, 2008).

### 2.2.8.4   Hybrid fabrics

The term hybrid refers to a fabric that has more than one type of structural fiber in its construction. In a multi-layer laminate, if the properties of more than one type of fiber were required, then it would be possible to provide this with two fabrics, each ply containing the fiber type needed. However, if low weight or extremely thin laminates are required, a hybrid fabric will allow the two fibers to be incorporated in just one layer of fabric instead of two. It would be possible in a woven hybrid to have one fiber running in the weft direction and the second fiber running in the warp direction, but it is more

*Table 2.8* Comparison of properties of common weave styles

| Property | Plain | Twill | Satin | Basket | Leno | Mock leno |
|---|---|---|---|---|---|---|
| Stability | ***** | *** | ** | ** | ***** | *** |
| Drape | ** | **** | ***** | *** | * | ** |
| Lack of porosity | *** | **** | ***** | ** | * | *** |
| Smoothness | ** | *** | ***** | ** | * | ** |
| Balance | **** | **** | ** | **** | ** | **** |
| Symmetry | ***** | *** | * | *** | * | **** |
| Lack of crimp | ** | *** | ***** | ** | ***** | ** |

Source: Gurit Composite Technologies, 2008.

Notes: *, very poor; **, poor; ***, acceptable; ****, good; *****, excellent.

*Figure 2.15* Hybrid fabric (twill weave) of aramid and carbon (Source: Fibre Glast Developments Corp., 2003).

common to find alternating threads of each fiber in each warp/weft direction. For instance, in an attempt to harness the stiffness and compressive strength of carbon fiber with the impact resistance and tensile strength of aramid fiber, engineers developed a hybrid fabric consisting of carbon and aramid as shown in Figure 2.15.

Although hybrids are most commonly found in 0/90° woven fabrics, the principle is also used in 0/90° stitched, unidirectional, and multi-axial fabrics. The most common hybrid combinations are as follows:

- Carbon/aramid: The high impact resistance and high tensile strength of the aramid fiber combines with the high compressive and tensile strengths of carbon. Both fibers have low density but relatively high cost.
- Aramid/glass: The low density, high impact resistance, and tensile strength of aramid fiber combines with the good compressive and tensile strength of glass, coupled with its lower cost.
- Carbon/glass: Carbon fiber contributes high tensile and high compressive strengths, high stiffness, and reduces the density, while glass reduces the cost (Gurit Composite Technologies, 2008).

### 2.2.8.5 Multi-axial fabrics

Multi-axial fabrics, also known as non-woven, non-crimped, stitched, or knitted, have optimized strength properties because of the fiber architecture. Stitched fabrics consist of several layers of unidirectional fiber bundles held together by a nonstructural stitching thread, usually polyester. The fibers in each layer can be oriented along any combination of axes between 0 and 90°. Multiple orientations of fiber layers provide a quasi-isotropic reinforcement. The entire fabric may be made of a single material, or different materials can be used in each layer. A layer of mat may also be incorporated into the

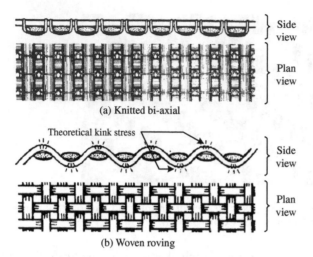

Figure 2.16 (a) Knitted biaxial fabric (b) Woven roving (Source: Greene, 1999).

construction. A schematic drawing of a typical knitted biaxial fabric is shown in Figure 2.16(a).

Conventional woven fabrics are made by weaving fibers in two perpendicular directions (warp and fill). However, weaving bends the fibers, reducing the maximum strength and stiffness that can be achieved. In addition, fabrics also tend to fray when cut, making them difficult to handle. Stitched fabrics offer several advantages over conventional woven fabrics. In the simplest case, woven fabrics can be replaced by stitched fabrics, maintaining the same fiber count and orientation. When compared to traditional woven fabrics, stitched fabrics offer mechanical performance increases of up to 20% over woven fabrics primarily from the fact that the fibers are always parallel and non-crimped, and that more orientations of fiber are available from the increased number of layers of fabric. Other noteworthy advantages of stitched fabrics include the following:

- Stress points located at the intersection of warp and fill fibers in woven fabrics are no longer present in stitched fabrics.
- Higher density of fiber can be packed into a laminate compared with a woven, essentially behaving more like layers of unidirectional.
- Heavy fabrics can be easily produced.
- Increased packing of the fiber can reduce the quantity of resin required (Gurit Composite Technologies, 2008).

Multiaxial fabrics have several disadvantages. First, the polyester fiber used for stitching does not bond very well to some resin systems and so the stitching can serve as a site for failure initiation. The production process can be quite slow and the cost of the machinery high. Consequently, stitched fabrics can be relatively expensive compared to woven fabrics. Extremely heavyweight fabrics can also be difficult to impregnate with resin without some automated process. Finally, the stitching process can bunch together the fibers, particularly in the 0° direction, creating resin-rich areas in the laminate.

For over half a century, these stitched fabrics have been traditionally used in boat hulls. Other applications include wind turbine blades, light poles, trucks, buses, and underground storage tanks. Currently, these fabrics are used in bridge decks and column repair projects. Woven and knitted textile fabrics are designated in ounces per square yard (oz/yd$^2$) (Gurit Composite Technologies, 2008).

### 2.2.8.6 *Woven roving*

Woven roving reinforcement consists of flattened bundles of continuous strands in a plain weave pattern with slightly more material in the warp direction. To form the material, roving is woven into a coarse, square, lattice-type, open weave as shown in Figure 2.16(b). Woven roving provides great tensile and flexural strengths and a fast laminate build-up at a reasonable cost. Woven roving is more difficult to wet-out than chopped strand mat however, and because of the coarse weave, it is not used where surface appearance is important. When more than one layer is required, a layer of chopped strand mat is often used between each layer of roving to fill the coarse weave. Woven roving is categorized by weight per unit area (oz/yd$^2$) and is sold by the running yard or in bulk by the pound or kilogram (Coast Fiber-Tek Products Ltd., 2003).

### 2.2.8.7 *Other forms*

A number of other forms of reinforcement are currently manufactured in the composites industry. Spun roving, for instance, is a heavy low-cost glass or aramid fiber strand consisting of filaments that are continuous but doubled back on themselves. Braids are fibers that are woven into a tubular shape instead of a flat fabric, as for a carbon fiber reinforced golf club shaft (Pebly, 1987). Some forms of reinforcement discussed earlier, as well as other types, are illustrated in Figure 2.17.

## 2.3 Matrix types

The primary functions of the matrix (or resin) in a composite are:

* to transfer stresses between fibers;
* to provide a barrier against the environment;
* to protect the surface of the fibers from mechanical abrasion.

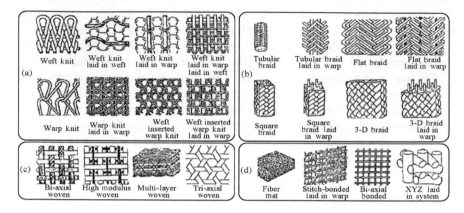

*Figure 2.17* Various reinforcement architectures including (a) Knits; (b) Braids; (c) Weaves; (d) Other forms (Source: Adopted from Ko, 2003).

The matrix plays a major role in a composite and influences the inter-laminar shear as well as the in-plane shear properties of the material. The interaction between fibers and matrix is important when designing damage-tolerant structures. Furthermore, the ability to manufacture the composite, and defects within it, depend strongly on the matrix's physical and thermal characteristics such as viscosity, melting point, and curing temperature (Mallick, 1993). There are generally two types of matrix: organic and inorganic.

### 2.3.1   Organic matrices

Organic matrices, also known as resins or polymers, are the most common and widespread matrices used today. All polymers are composed of long chain-like molecules consisting of many simple repeating units. Polymers can be classified under two types – thermoplastic and thermosetting – according to the effect of heat on their properties. Like metals, thermoplastics soften with heating and eventually melt, hardening again with cooling. This process of crossing the softening or melting point can be repeated as often as desired without any noticeable effect on the material properties in either state. Typical thermoplastics include nylon, polypropylene, polycarbonate, and polyether-ether ketone (PEEK).

Thermosets are formed from a chemical reaction when the resin and hardener (or catalyst) are mixed and then undergo a nonreversible chemical reaction to form a hard, infusible product. In some thermosets, such as phenolic resins, volatile by-products are produced. Other thermosetting resins, such as polyester and epoxy, cure by mechanisms that do not produce any volatile by-products and, thus, are much easier to process. Once cured,

*Table 2.9* Comparison of resins

| Resin type | Advantages | Disadvantages |
|---|---|---|
| Polyesters | Easy to use<br>Lowest cost of resins available | Only moderate mechanical properties<br>High styrene emissions in open molds<br>High cure shrinkage<br>Limited range of working times |
| Vinyl esters | Very high chemical/<br>environmental resistance<br>Higher mechanical properties<br>than polyesters<br>High mechanical and thermal<br>properties | Post-cure generally required for high<br>properties<br>High styrene content<br>Higher cost than polyesters<br>High cure shrinkage |
| Epoxies | High water resistance<br>Long working times available<br>Temperature resistance up to<br>140 °C wet / 220 °C dry<br>Low cure shrinkage | More expensive than vinyl esters<br>Corrosive handling<br>Critical mixing |

Source: Gurit Composite Technologies, 2008.

thermosets will not become liquid again if heated, although above a certain temperature their mechanical properties will change significantly (Mallick, 1993; Gurit Composite Technologies, 2008).

In general, the three most common of organic resins currently used are polyester, vinyl ester, and epoxy. A brief description of each resin is presented in Table 2.9 and in the following sections.

### 2.3.1.1 Polyester

Polyester resins are the most economical and widely used resin systems, especially in the marine industry. Nearly half a million tons of this material are used annually in the United States in composite applications. Polyester resins can be formulated to obtain a wide range of properties ranging from soft and ductile to hard and brittle. Their advantages include low viscosity, low cost, and fast cure time. In addition, polyester resins have long been considered the least toxic thermoset resin. The most significant disadvantage of polyesters is their high volumetric shrinkage (Mallick, 1993; Gurit Composite Technologies, 2008).

### 2.3.1.2 Vinyl ester

Vinyl ester resins are more flexible and have higher fracture toughness than cured polyester resins. The handling and performance characteristics of vinyl esters are similar to polyesters. Some advantages of the vinyl esters, which may justify their higher cost, include better chemical and corrosion resistance, hydrolytic stability, and better physical properties, such as tensile strength as well as impact and fatigue resistance. It has been shown that a 0.02 to 0.06 in. (0.5 to 1.5 mm) thick layer of a vinyl ester resin matrix can provide an excellent permeation barrier to resist blistering in marine laminates (Mallick, 1993).

### 2.3.1.3  Epoxy

Epoxy resins are a broad family of materials that provide better performance as compared to other organic resins. Aerospace applications use epoxy resins almost exclusively, except when high temperature performance is a key factor. Epoxies generally out-perform most other resin types in terms of mechanical properties and resistance to environmental degradation. The primary advantages of epoxy resins include:

- wide range of material properties;
- minimum or no volatile emissions and low shrinkage during cure;
- excellent resistance to chemical degradation;
- very good adhesion to a wide range of fibers and fillers.

The high cost of epoxies, long cure time, and handling difficulties are the principal disadvantages (Mallick, 1993).

## References

Advanced Composites, Inc. (2003) *Custom Products and Technological Capabilities.* [On-line] Available from: www.advancedcomposites.com. [Accessed 1 August 2003].

Albarrie Canada Limited (2003) *Basalt Product Information.* Ontario, Canada.

Amateau, M.F. (2003) *Properties of Fibers.* Course Notes for Mechanical Engineering 471: Pennsylvania State University, Engineering Composite Materials.

Bascom, W.D. (1987) Fiber sizing. *Engineered Materials Handbook – Volume 1: Composites.* Metals Park, OH: American Society of Metals.

Celanese Acetate LLC (2001) *Complete Textile Glossary.* New York.

Coast Fiber-Tek Products Ltd (2003) *Fiberglass Materials.* [On-line] Available from: www.fiber-tek.com. [Accessed 1 August 2003].

Competitive Cyclist (2003) [On-line] Available from: www.competitivecyclist.com [Accessed 1 August 2003].

Composite Basics (2003) [On-line] Available from: www.compositesone.com. [Accessed 1 August 2003].

Composite Tek (2003) *Selecting the Right Fiber: The Lightweight, High Strength and Stiffness Solution.* Boulder, CO: Resources & Helpful Hints.

Cornelius, C. and Marand, E. (2002) Hybrid inorganic–organic materials based on a 6FDA–6FpDA–DABA polyimide and silica: physical characterization studies. *Polymer,* 43, pp. 2385–2400.

Cumming, W.D. (1987) Unidirectional and two-directional fabrics. *Engineered Materials Handbook – Volume 1: Composites.* Metals Park, OH: American Society of Metals.

Department of Defense (1996) *Polymer Matrix Composites.* MIL-HDBK-17-3E, DOD Coordination Working Draft.

Dominguez, F.S. (1987) Woven fabric pre-pregs. *Engineered Materials Handbook – Volume 1: Composites*. Metals Park, OH: American Society of Metals.

Fibre Glast Developments Corp. (2003) *Fiberglass Fabrics Products Catalog*. Ohio.

Greene, E. (1999) *Marine Composites*. Annapolis, MD: Eric Green Associates, Inc.

Gurit Composite Technologies (2008) *Guide to Composites*. [CD-ROM]. Gurit, Switzerland: Zurich. www.gurit.com.

Hancox, N.L. (1981) *Fibre Composite Hybrid Materials*. New York: MacMillan Publishing Co.

Hansen, N.W. (1987) Carbon fibers. *Engineered Materials Handbook – Volume 1: Composites*. Metals Park, OH: American Society of Metals.

Ko, F.K. (2003) *Textile Structural Composites*. Beijing, China: Scientific Publishing Co.

Mallick, P.K. (1993) *Fiber-Reinforced Composites: Materials, Manufacturing, and Design*. New York: Marcel Dekker, Inc., 566 pp.

Miller, D.M. (1987) Glass fibers. *Engineered Materials Handbook – Volume 1: Composites*. Metals Park, OH: American Society of Metals.

Moy, S. (2002) *Design Guidelines: Partial Safety Factors for Strengthening of Metallic Structures*. Presentation, University of Southampton. [On-line] Available from: www.cosacnet.soton.ac.uk/presentations/5thMeet/moy_5th.pdf

Pebly, H. (1987) Glossary of terms. *Engineered Materials Handbook – Volume 1: Composites*. Metals Park, OH: American Society of Metals.

Pilling, J. (2008) *Composite Materials Design*. [On-line], Available at: www.callisto.my.mtu.edu/my4150/ Michigan Technological University, [Accessed August 18, 2008].

Plant of Insulation Materials, Ltd. (2003) *Materials*. [On-line] Available from: www.basaltfibre.com/eng/

Saint Gobain Corporation (2003) *Technical Data Sheet: 952 Chopped Strand for Thermoplastic Polyester and PC Resins*. Canada: Saint Gobain Corporation.

Schwartz, M.M. (ed.) (1985) *Fabrication of Composite Materials Source Book*. Metals Park, OH: American Society for Metals.

Shan, Y. and Liao, K. (2002) Environmental fatigue behavior and life prediction of unidirectional glass–carbon/epoxy hybrid composites. *International Journal of Fatigue*, 24(8), pp. 847–859.

Smith, Z. (1996) *Understanding Aircraft Composite Construction*. NAPA, CA: Aeronaut Press.

Watson, J. and Raghupathi, N. (1987) Glass fibers. *Engineered Materials Handbook – Volume 1: Composites*. Metals Park, OH: American Society of Metals, pp. 107–111.

Zoltek Corporation. (2008) *Zoltek Panex Carbon Brochure*. Missouri: Zoltek Corp.

# 3 Fabrication techniques

## 3.1 Introduction

It is well known that fiber and matrix type largely influence the overall mechanical properties of a composite. However, the end properties of a composite produced from these materials are also a function of the way in which the materials are prepared and processed. For most commercial repair systems available for civil infrastructure applications, the manufacturers provide both the fibers in the sheet or fabric form and the matrix. The fibers are normally treated with coatings that provide good wetting for the chosen matrix. They also provide polymers for pre-treatment of surfaces and cover coats. In most cases, the sheets or fabrics are saturated with the matrix and applied to the prepared surface (manual lay-up), then allowed to cure. Preparation includes cleaning, filling of voids and application of prime coat on the parent material (substrate). After the matrix cures, a topcoat is applied for protection and aesthetics. In some cases, fireproof coatings are used to improve the fire resistance of the repair. The composites are also available in preformed, pre-cured sheets. These laminates, made of unidirectional carbon fibers, have much higher tensile strength because of better alignment and factory processing. These thin laminates are attached to the concrete surface using polymers. Here again, the manufacturers supply compatible primer and adhesives.

This chapter presents the methods of preparing laminate samples, including hand lay-up technique, vacuum bagging, and pultrusion.

## 3.2 Manufacturing methods

### 3.2.1 Fiber impregnation

In many cases, FRP composites are constructed manually using a hand-impregnation technique. Also referred to as hand lay-up, this process involves placing (and working) successive plies of resin-impregnated reinforcement in position by hand. Squeegees, brushes, and grooved rollers are used to force the resin into the fabric and to remove much of the entrapped air. This method has become commonplace throughout the composites industry for a number of reasons. First, the principles of this technique are easy to teach and have been used widely for many years. Second, this method is very economical since large expensive equipment is not needed. Last, a wide variety of fibers and resins are compatible with this method and can be purchased through a large number of suppliers.

Unfortunately, this method can lend itself to a host of problems, especially if air voids remain within the composite. These air voids can eventually form cracks, which can propagate throughout the composite structure. This will result in a delamination or debonding failure in which the bond between FRP layers breaks down, allowing the composite layers to separate. Not only will this lead to a significant reduction in strength, but it will also allow adverse environmental conditions to penetrate and attack the surface of the FRP layers. Other noteworthy disadvantages include the following:

- Resin mixing, laminate resin content, and overall laminate quality are strongly dependent upon the skills of the laborers.
- Hand lay-up is a very slow and labor-intensive process.
- Health and safety risks may be posed when handling resins. Since the fabric must be impregnated by hand, the resins have a low viscosity to facilitate easier wet-out of the fabric. Unfortunately, the lower viscosity of the resins also means that they are more like to penetrate clothing and harm the skin. In addition, lower viscosity resins generally have lower mechanical and thermal properties (Mallick, 1993; May, 1987; Pebly, 1987; SP Systems, 2001).

### 3.2.2  Vacuum bagging system

When utilizing FRP material systems in a design, it is usually assumed that the materials are joined together as a unified structure. Strength and stiffness predictions rely solely on the assumption that constituents are completely bonded together to form one cohesive element. Therefore, the bond between polymer matrix and the fiber reinforcement is critical in determining the mechanical properties of the resultant composite. Full impregnation is the most critical to achieve good adhesion and strong bonding. During impregnation, all reinforcement surfaces must be exposed to the resin, otherwise, gas-filled bubbles, air voids, crevices, and other discontinuities or defects will remain, adversely affecting the mechanical properties of the finished composite. For example, among the individual fiber filaments are microscopic pores and interstices filled with air. This air may prevent the resin from fully impregnating some of these capillary type passages, and sufficient wetting of their surfaces will not occur, resulting in poor bonding. Consequently, these air voids will be primary sites for failure initiation and crack propagation. Any area of the structure not fully bonded is considered a location for potential failure or at the very least is not performing to its full potential. This is neither an economical use of materials nor an efficient building practice (Diab, 2001; May, 1987).

In addition to structural problems, entrapped air may also lead to cosmetic problems. As the resin cures, gases may be released and accumulate in voids in the form of a blister. If the composite is exposed to sunlight or ultraviolet rays, this process will be aggravated. In addition, voids close to the surface of the composite may absorb moisture or collect dirt or debris, causing unsightly blemishes on the surface of the composite material (Diab, 2001).

To overcome many of these problems, a denser FRP must be manufactured by removing nearly all air voids within the composite. The simplest way to remove air voids and develop a strong bond is to apply pressure to the wet composite during lay-up. Early manufacturers of composites stacked lead weights or bricks on top of the composite. Unfortunately, these concentrated loads produced spotty bonds and forced resin out from underneath the weights, leaving a resin-starved area in the composite. In addition, areas with less weight tend to pool resin and be much heavier.

Two alternative methods, which are capable of applying pressure more efficiently, are vacuum bagging and pressure bag molding (pressure bagging). Vacuum bagging is an economical and effective method that has been used universally for the manufacture of aerospace structures. In this process, fiber reinforcement layers are first impregnated with resin, then stacked together and placed inside a sealed bagging system, as shown in Figure 3.1.

A vacuum pump is then attached to the bag, removing the air within the bag and allowing external atmospheric pressure to firmly press the FRP composite. The wet FRP layers are pressed tightly against the surface being covered so that the excess resin is squeezed out and soaked up in a disposable outer wrap. The vacuum-bagging system allows for predictable and consistent pressure application, providing control on FRP thickness, reducing void content, improving resin flow, and assisting in bonding. Pressure bags also invoke the use of pressure but are considerably more complex and expensive to operate. They apply additional pressure to the assembly through an elastomeric

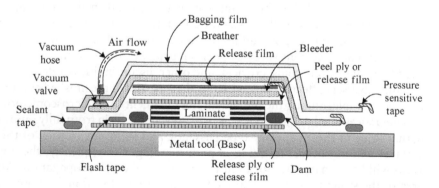

*Figure 3.1* Vacuum bagging setup for laminate composites (Source: Richmond Aircraft, 2008).

pressure bag or bladder contained within a clamshell cover, which fits over a mold. However, only mild pressures can be applied with this system (May, 1987). Since simplicity is often desired in nearly all FRP applications, the more suitable method is vacuum bagging. The most critical element of a vacuum bagging system is that a smooth surface must be provided around the perimeter of the bag to create an airtight seal (Diab, 2001).

Vacuum bagging provides both vacuum as well as pressure. The vacuum is responsible for drawing out volatiles and trapped air, resulting in a low void content. Pressure compacts the laminate, providing good consolidation and interlaminar bonds. Regardless of whether the part is made up of vertical, horizontal, curved, compound curved surfaces, or a combination of these, the same amount of pressure is applied everywhere. This even distribution of vacuum pressure results in improved control of the volume fraction of fiber, an important design parameter related to FRP laminate strength (Diab, 2001).

### 3.2.3 Pultrusion

Pultrusion is an automated, continuous process for manufacturing structural shapes with constant cross-section. In this method, dry reinforcing fibers are pulled through a resin, usually followed by a pre-forming system, then into a heated die, where the resin undergoes polymerization and curing (Figure 3.2). Pultrusion is a unique process in that the incoming fibers are usually unidirectional and must be pulled through the pultrusion die since the uncured material lacks the rigidity to be pushed through the die, as in an extrusion process, where a billet of raw material is pushed through a die. The pultrusion

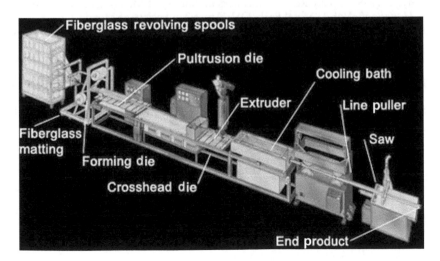

*Figure 3.2* Schematic of pultrusion manufacturing process. (Source: Performance Fiberglass Inc., 2008).

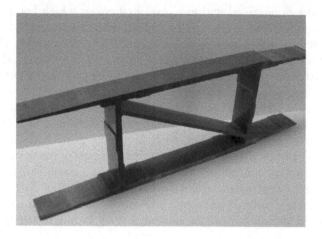

*Figure 3.3* Typical beam cross-section manufactured using pultrusion process.

process generates long lengths of material with high unidirectional strengths (Figure 3.3) (Jones, 1999).

## References

Diab Technologies (2001) *Advanced Vacuum Bagging Techniques on Sandwich Constructions: Technical Bulletin.* DeSoto, TX: Diab Group.
Jones, R.M. (1999) *Mechanics of Composite Materials.* 2nd edn, New York & London: Brunner-Routledge.
Mallick, P.K. (1993) *Fiber-Reinforced Composites: Materials, Manufacturing, and Design.* New York: Marcel Dekker, Inc.
May, C. (1987) Epoxy Resins. In: American Society of Metals, *Engineered Materials Handbook,* Vol. 1, Metal Park, OH: American Society of Metals.
Pebly, H (1987) Glossary of Terms. In: American Society of Metals, *Engineered Materials Handbook,* Vol. 1, Metal Park, OH: American Society of Metals.
Performance Fiberglass Inc., *The Fiberglass Story.* [Online image] Available at: <http://www.performancefiber.com> [Accessed 15 August 2008].
Richmond Aircraft Products (2008) *Product Line Technical Data.* California: Umeco Composites.
SP Systems: Guide to Composites (2001) [CD-ROM]. *Structural Polymer Systems Limited.* Isle of Wight, UK, 69 pages.

# 4 Common repair systems

## 4.1 Introduction

Externally bonded steel plates, steel or reinforced concrete jackets, and external prestressing were the traditional techniques used to strengthen or retrofit existing concrete structures to provide additional strength, compensate for construction design errors, correct deterioration problems, or improve the performance as required by code changes. These techniques were expensive because of the heavy equipment needed and, in most instances, did not provide a long service life. Corrosion of reinforcement, which is the common deterioration mechanism, affects the life cycle of steel plate in a similar manner. In addition, in certain cases such as strengthening of unreinforced masonry and concrete walls, the traditional techniques cannot be used. High-strength composite materials, used by the aerospace industry for more than 40 years, are becoming the preferred materials for the repairs. Since composite materials are extremely light, the process of installing the composite onto a structure is simplified considerably and, in most cases, heavy equipment is not needed for the repairs. The composites also have very high corrosion resistance. Although the materials are expensive, the final cost of repair becomes competitive because of the ease of construction and the time savings. The most popular uses include:

- strengthening of reinforced and prestressed concrete beams for flexure;
- shear strengthening of reinforced and prestressed concrete beams;
- column wrapping to improve the ductility for earthquake-type loading;
- strengthening of unreinforced masonry walls for in-plane and out-of-plane loading;
- strengthening for improved blast resistance.

Philosophically, the fiber-reinforced polymer (FRP) bonded to the parent structural element acts as reinforcement. Although research has been carried out to evaluate the use of prestressed FRP, almost all applications involve the attachment of unstressed FRP to the parent structures using epoxy. The FRP essentially acts as additional reinforcement. The major differences, as compared to traditional reinforcing bars, are (i) the strength of FRP is much higher; (ii) the behavior of most FRP is linearly elastic up to failure; (iii) the load transfer from FRP to concrete is by adhesion of epoxy as opposed to mechanical bond between bars and concrete; and (iv) the structural elements are under stress (load) when FRP is applied and these stresses and corresponding strains should be taken into account in the design.

## 4.2   Fiber reinforced polymers: constituent materials and properties

This section provides a brief description of the two basic constituents of FRP, namely, fibers and matrix. More details can be found in Chapter 2. The popular fibers are made of aramid (Kevlar®), carbon, or glass. The fibers are available in various forms including tows (roving) that consist of 1000, 12,000, 48,000 fibers, unidirectional sheets, and woven fabrics. Tows with more than 48,000 fibers are also available.

Aramid fibers can be grouped into general-purpose or high-performance fibers. Both groups have a tensile strength in the range of 493–595 ksi (3.4–4.1 GPa). The general-purpose fibers have a modulus of 10,000–12,000 ksi (69–83 GPa), and a rupture strain of 2.5%, whereas the high-performance fibers have a modulus of 15,900–18,000 ksi (110–124 GPa) with a rupture strain of 1.6%. The fiber content in aramid composite ranges from 40 to 60%. The density of aramid fiber is about 90.5 lb/ft$^3$ (1450 kg/m$^3$). Longitudinal thermal expansion (in the fiber direction) is about $-2 \times 10^{-6}$/°C. These fibers are susceptible to fire. Typical commercial fiber types are Kevlar®49 and Twaron1055.

Carbon fibers are the most popular fibers for composites because of their high strength and excellent durability. The various types of commercially available fibers can be grouped as: general-purpose fibers with a modulus of 31,908–34,809 ksi (220–240 GPa) and a strength of 290–551 ksi (2.0–3.8 GPa), high-strength fibers with a strength of 551–696 ksi (3.8–4.8 GPa), ultra high-strength fibers with a strength of 696–899 ksi (4.8–6.2 GPa), high-modulus fibers with a modulus of 49,312–75,419 ksi (340–520 GPa) and a strength of 246.6–464 ksi (1.7–3.2 GPa), and ultra-high-modulus fibers with a modulus of 75,419–100,076 ksi (520–690 GPa) and a strength of 203–348 ksi (1.4–2.4 GPa). The density of carbon fibers range from 96.7 to 106 lb/ft$^3$ (1550 to 1700 kg/m$^3$). The thermal expansion coefficient ranges from $-0.1$ to $-0.5 \times 10^{-6}$/C.

A number of formulations are available for glass fibers. The most economical fibers are E-glass fibers with a modulus of 10,152.6 ksi (70 GPa) and strength of about 290 ksi (2 GPa). Stronger S-glass fibers can sustain stresses up to 580 ksi (4 GPa) and their modulus is about 20% higher than E-glass fibers. Alkali-resistant or AR-glass fibers have properties similar to E-glass fibers but are more durable in alkali environment such as concrete.

The density of glass fibers is about $156 \, lb/ft^3$ ($2500 \, kg/m^3$). The thermal expansion coefficient varies from 3 to $5 \times 10^{-6}/°C$.

In almost all field applications, the matrix consists of thermosetting polymers (epoxies, polymers, or vinyl esters). Unsaturated polyester is the most popular product in North America. Typical tensile strength and modulus are 10.8 ksi (75 MPa) and 478.6 ksi (3.3 GPa), respectively. Epoxy resins are typically more expensive than commercial polyesters and vinyl esters, and most epoxies provide better mechanical properties if cured at a higher temperature.

For most commercial systems available for civil infrastructure applications, the manufacturers provide both the fibers – in the sheet or fabric form – and the matrix. The fibers are normally treated with coatings that provide good wetting for the chosen matrix. They also provide polymers for pre-treatment of surfaces and cover coats. In most cases, the sheets or fabrics are saturated with the matrix, applied to the prepared surface, and allowed to cure. Preparation includes cleaning, filling of voids, and application of the prime coat. After the matrix cures, a topcoat is applied for protection and aesthetics. In some cases, fireproof coatings are used to improve the fire resistance of the repair. The composites are also available in preformed sheets. These plates, made of unidirectional carbon fibers, have much higher tensile strength because of better alignment and factory processing. These thin plates are attached to the concrete surface using polymers. Here again, the manufacturers supply compatible primer and adhesives.

Fiber volume fraction for factory-produced composites varies from 60 to 70%. For composites fabricated in the field, the fiber volume fractions typically range from 25 to 40%. Depending on the quality of fibers and methods used for repairs, the tensile strength of composites made using unidirectional carbon fibers could range from 116 to 145 ksi (0.8 to 1.0 GPa). For factory-produced sheets, the strength could be as high as 507.6 ksi (3.5 GPa). For glass fiber composites, the tensile strength for unidirectional sheets range from 72 to 203 ksi (0.5 to 1.4 GPa) and the range for aramid fibers is 101.5–246.6 ksi (0.7–1.7 GPa). The stress–strain behavior for both fiber and the composites is linear up to failure (SP Systems, 2001).

## 4.3 FRP for increasing flexural strength

FRP can be effectively used to increase the flexural strength by attaching unidirectional sheets at the extreme tension surface. Although strength increases up to 160% have been reported in the literature (Meier and Kaiser, 1991; Ritchie *et al.*, 1991), ductility and serviceability constraints limit the percentage increase to about 40%. Typical load–deflection responses of control and strengthened beams are shown in Figure 4.1. It can be seen in the figure that FRP increases both the maximum load and the stiffness, and there is a noticeable reduction in deflection at failure. In most cases, beams strengthened with organic polymers fail by delamination of the FRP plate. In certain cases,

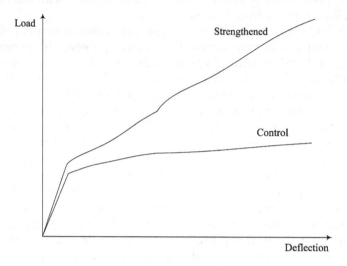

*Figure 4.1* Load versus deflection curves for both control and strengthened beams.

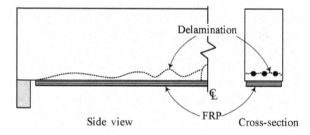

*Figure 4.2* Delamination at the reinforcement level (Source: ACI Committee 440, 2002).

failure occurs due to tension failure of the concrete cover. When this happens, the delamination occurs at the original reinforcement (bar) level as shown in Figure 4.2. These failure modes limit the number of layers (or the area of FRP) that can be applied and, hence, the strength increases. Additional information regarding the failure modes and discussion on strength increases can be found in the literature (Arduini and Nanni, 1997; Nakamura *et al.*, 1996; Ross *et al.*, 1999; Saadatmanesh and Ehsani, 1991; Sharif *et al.*, 1994).

### 4.3.1  *Analysis and design*

For design purposes, the FRP is assumed to act as additional reinforcement. Depending on the dimensions of the beam, loading conditions, amount of steel reinforcement, and the area of FRP, the possible failure modes are as follows:

1. Crushing of concrete before yielding of steel; this failure occurs because of over-reinforcement. Using American Concrete Institute (ACI) Code recommendations, crushing of concrete can be assumed when the maximum compressive strain reaches 0.003 (ACI Committee 318, 2005).
2. Yielding of the steel in tension followed by rupture of FRP laminate; this failure allows for the full utilization of mild steel reinforcement and provides some ductility. Even after yielding of steel, the load–deflection curve will have substantial positive slope or flexural rigidity because of the contribution of the FRP reinforcement. Since the FRP behavior is linearly elastic to under-reinforced, the failure will be brittle as compared to typical under-reinforced RC beams.
3. Yielding of steel in tension followed by crushing of concrete; this failure occurs when the area of FRP is sufficient to generate more tension force than the concrete can generate in compression after yielding of steel. In this situation, the under-reinforced beam is transformed to an over-reinforced beam.
4. Shear/tension delamination of the concrete cover (Figure 4.2) or debonding of the FRP from concrete substrate; although models are available to estimate flexural capacities for these failure modes, it is not recommended to design beams using this failure model because the failure mode is very brittle. This failure mode can be prevented by limiting the maximum stress allowed on the FRP, especially for multiple layers.

### 4.3.2 Flexural strengthening systems design

The major steps for designing the strengthening system are as follows:

1. Obtain the details of the current slab or beam. These include the compressive strength of concrete, location and yield strength of steel and shear reinforcement.
2. Estimate the loading that will be present during the application of the strengthening system. Every effort should be made to minimize the superimposed load that will be present during the application of strengthening. Reduction in superimposed load during repair increases the efficiency of the repair.
3. Choose the strengthening system. This could be carbon plate, carbon fiber sheets applied in the field, or glass fiber sheets applied in the field.
4. Obtain the geometric and mechanical properties of the chosen fiber composite system. This information can be obtained from the manufacturers.
5. Estimate the moment capacity of the existing beam (slab).
6. Estimate the moment capacity needed for upgraded loads.
7. Estimate the amount of composite reinforcement needed. The strength computation equations will provide an approximate area. This area should be converted into plate thickness and width or number of layers

and width of sheets (fabrics). This step can be considered as preliminary design. The composite should generate sufficient additional moment capacity to resist the extra loads. In some instances, composite reinforcement can be provided to correct the deficiencies such as reinforcement lost to corrosion.

8. Check to ascertain that the working load moment for the upgraded loads do not exceed the following stresses:

- Maximum concrete stress $<0.45f'_c$, in which $f'_c$ is the compressive strength of concrete.
- Maximum steel stress in tension $<0.8f_y$, in which $f_y$ is the yield strength of steel.
- Maximum steel stress in compression $<0.4f_y$.
- Maximum composite stress $<0.33f_u$, in which $f_u$ is the fracture strength of the composite. Depending on the type and exposure, the allowable composite stresses might have to be reduced further.

9. Check for ductility requirements.
10. Check for creep-rupture and fatigue stress limits.
11. Check for deflection limits.

### 4.3.3  Preliminary design (estimation of composite area)

An initial estimate of the composite area can be made using the following guidelines. Once the area of composite is established, detailed analysis should be carried out to check all the requirements at working and ultimate loads.

1. Compute the design ultimate moment, $M_u$, based on the upgraded loads.
2. Compute the factored nominal moment capacity, $\phi M_{ni}$, of unstrengthened cross-section. The strength reduction factor, $\phi$, can be assumed as 0.90.
3. Assuming a failure strain of about $0.8\varepsilon_{fu}$ for composite, and a lever arm of $0.9h$, estimate the area of composite, $A_f$. It is assumed that the composite will be attached at the extreme tension face and the net fiber area will be used for computations.
4. Decide on the width of the composite. Normally, the width of the beam will control the width of the composite. In the case of slabs, a 2.95 or 3.94 in. (75 or 100 mm) wide strip could be used for each 11.8 in. (300 mm) width of slab.
5. Estimate the thickness of the plate or the number of layers of sheet reinforcement. The width and number of layers can be adjusted simultaneously to obtain round numbers. In the case of multiple layer application, it is advisable to use the same width for all layers.

### 4.3.4  Final design

Once the area of composite (or fiber area) is determined, the properties of the entire section are known for checking the various requirements. The following sequence of calculations can be used as a guideline.

1.  For the unstrengthened section, obtain the properties of uncracked section and cracking moment. These include computation of Young's modulus of elasticity of concrete, $E_c$, depth of neutral axis of uncracked section, $\bar{y}$, moment of inertia of uncracked section, $I_g$, and cracking moment, $M_{cr}$.
2.  Compute the properties of cracked, unstrengthened section. These include $kd$ and $I_{cr}$.
3.  For the loads present at the installation of composite, compute maximum stresses in concrete and steel, $f_c$, $f_s$, $\varepsilon_{bi}$ (extreme tension fiber), and maximum deflection, $\delta$. Note that for this step, the properties of the unstrengthened section should be used.
4.  Compute $\phi M_n$ for the strengthened section. If $\phi M_n \geq M_u$, proceed further. Otherwise, revise the area of composite and recompute $\phi M_n$.
5.  Check for ductility requirements.
6.  Compute $f_c$, $f_s$, and $f_f$ for the revised loads. Note that the stresses for the difference between the original and upgraded loads (moment) should be computed using the properties of the strengthened section and added to the stresses obtained in Step 3.
7.  If allowable stresses for worked load are satisfied, proceed to Step 8. Otherwise, revise the area of composite and repeat Steps 4, 5, and 6, or just 6.
8.  Compute the deflection and check for allowable limits.
9.  Check for other strength parameters such as shear.

More details on the various limits and a design example can be found in the ACI Committee 440 Report, "Guide for the Design and Construction of Externally Bonded FRP Systems for Strengthening Concrete Structures," published by the American Concrete Institute (2002).

## 4.4  Strengthening for shear

FRP systems have been found effective for increasing the shear capacity of beams (Chajes *et al.*, 1995; Kachlakev and McCurry, 2000; Malvar *et al.*, 1995; Norris *et al.*, 1997). Conceptually, the fibers in FRP should be placed perpendicular to the potential shear cracks. FRP is wrapped around the web of the beams, using complete or three-sided U-wraps as shown in Figure 4.3. FRP can also be placed only on sides, but this scheme is not recommended because FRP may not have enough bond area to develop full tensile strength

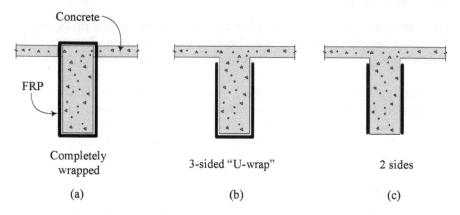

*Figure 4.3* Typical FRP shear reinforcement layout (Source: ACI Committee 440, 2002).

of the fibers. Along the span, the FRP system can be installed continuously covering the entire span or applied in discrete strips.

### 4.4.1   Analysis and design

As in the case of flexure, the contribution of composite shear reinforcement is added to the contributions of concrete and steel (shear) reinforcement. The contributions of concrete, $V_c$, and steel reinforcement, $V_s$, can be calculated using the procedures of reinforced concrete design. The shear contribution of composite reinforcement, $V_f$, is then added to the contributions of the concrete and steel reinforcement. Once the shear contribution composite, $V_f$, is established, the analysis and design can be done using the same procedure used for reinforced concrete. The major design steps are presented in the following section.

Allowable maximum strain, $\varepsilon_{fe}$, is a critical parameter. Since aggregate interlock provides some shear resistance, the maximum crack widths for shear cracks should be limited. To satisfy this condition, a maximum strain of 0.004 is recommended. The maximum strain should also not exceed 75% of the fracture strain of the fibers. This limit becomes critical when high-modulus carbon fibers are used for the shear strengthening. The third limitation is based on the tension force that can be generated before debonding. This limitation, which is applicable to two- and three-sided wraps, depends on the number of layers of fabric, thickness of the fabric, stiffness of the fibers, and the compressive strength of concrete (Khalifa *et al.*, 1999).

## 4.5 Axial compression, tension, and ductility enhancement

FRP can be used effectively to confine concrete, thereby increasing both the failure stress (compressive strength) and the corresponding strain. The increase in strain increases the ductility of the structural member and the system; hence, it can be used to improve the seismic performance. In fact, column wrapping to improve the seismic performance is one of the major practical application areas of FRP. The increases in both compressive strength and strain capacity, translated to curvature and rotation capacity, are very useful for columns subjected to dynamic loading (Restrepo and DeVino, 1996).

Bonded FRP can be used to increase the tensile capacity of concrete members. The contribution of FRP is limited by the force that can be transferred at the interface. Hence, the bond strength and bond area become very important. The principles used for FRP shear reinforcement for maximum strain are applicable for this case.

## 4.6 Recent advances

A number of advances have been made in the area of materials and design procedure. It is recommended that the reader seek the latest report from American Concrete Institute (ACI), Japan Concrete Institute (JCI), ISIS Canada, or CEB for the use of FRP. For example, ACI Committee 440 published a design guideline document 2002 (ACI Committee 440, 2002), and similar documents are under preparation. JCI also updates documents frequently. ISIS publications can be obtained from the University of Manitoba. Since this is an emerging technology, changes are being made frequently to design documents to incorporate recent findings. In the area of fibers, the major development is the reduction in the cost of carbon fibers. Other advances include the development of high-modulus (up to 690 GPa) carbon fibers and high-strength glass fibers.

## 4.7 Field applications

A large number of field applications have been carried out during the last 20 years. The majority of the initial uses were in Japan, followed by applications in Europe and North America. In North America, the popular applications are in rehabilitation of bridges to improve earthquake resistance, repair and rehabilitation of parking structures, strengthening of unreinforced walls, and rehabilitation of miscellaneous structures such as tunnels, chimneys, and industrial structures such as liquid retaining tanks. Typical examples are shown in Figures 4.4–4.6 (ACI Committee 440, 1996).

(a)

(b)

*Figure 4.4* Rehabilitation of parking garage with carbon FRP: (a) corrosion
prior to repair; (b) after FRP application (Source: Premier Corrosion
Protection – Tampa, 2008).

## 4.8   Construction procedures and quality assurance

The construction process is critical to obtaining the strength increases
provided by the FRP. Quality control is critical because the weakest location
controls the usable strength increase. Consequently, as FRP materials are
manufactured, a quality-control program must be implemented to assess and
monitor the properties of the material.

### 4.8.1 Witness panels

The quality-assurance program usually consists of fabricating "witness panels," which are small samples of a composite part, typically required when strengthening concrete structures with FRP materials. Two witness panels are typically required for each day of laminating, one panel fabricated near the beginning and one near the end of each workday. The panels are fabricated daily on the job site using the same materials prepared for application of the FRP to the structure, and under the same environmental conditions that are found on the job site. They are typically prepared on a smooth surface, such as a sheet of glass. The panels are typically a laminate of two or three stacked plies of identical fiber orientation and are, at a minimum, 11.8 × 11.8 in. (300 × 300 mm). Once laminated, the witness panels are allowed to cure on site for about 72 h before delivery to an accredited laboratory or test agency (Edge Structural Composites, 2003).

At the testing facility, test specimens are prepared using some of the panels. The tensile specimens are then tested to ensure that the materials applied on the job site meet, or exceed, the required properties and performance specified in the design. If the witness panel does not meet the specifications, additional layers may have to be applied, or the FRP may have to be removed and reapplied (Edge Structural Composites, 2003).

*Figure 4.5* Stiffening of cantilever slab with carbon FRP strips: (a) during retrofit; (b) after FRP application and sealant (Source: Premier Corrosion Protection – Tampa, 2008).

(b)

*Figure 4.5* (Continued).

(a)

*Figure 4.6* Strengthening of dock for additional load using carbon fiber: (a) during FRP application; (b) during application of sealant (Source: Premier Corrosion Protection – Tampa, 2008).

(b)

*Figure 4.6* (Continued).

## 4.8.2 Test method

The tensile samples are typically tested in accordance with ASTM D3039, "Standard Test Method for Properties of Polymer Matrix Composite Materials." In this method, the tensile specimen, which is rectangular in cross-section, is mounted in the self-aligning grips of a mechanical testing system and is subjected to monotonic tensile load. The ultimate strength of the composite is calculated using the maximum load while the stress–strain response is determined using strain or displacement transducers. If used, this equipment can provide information on other important properties, which include ultimate tensile strain, tensile modulus of elasticity, and Poisson's ratio (American Society for Testing and Materials, 2006).

This test could result in scatter within the data. The variability can be attributed to a number of factors such as material and specimen preparation, gripping, and system alignment. Inferior fabrication practices, poor fiber alignment, and damage incurred during coupon machining are problems encountered during preparation. Grip-induced failures are often due to poor tabbing or a lack of suitable reinforcement on the gripped ends of the specimen. Improper system alignment could result in bending of the specimen. Bending may occur due to misaligned grips or improper specimen placement in the grips. Bending will cause premature failure of the specimen as well as an inaccurate modulus of elasticity. Other sources of variability include testing temperature, void content, testing environment, and rate of loading (American Society for Testing and Materials, 2006).

### 4.8.3   Data sets

The intent of the data collection and analysis was simply to gather useful information on the statistical variability found in the test results of FRP tensile specimens, rather than to investigate the performance of the FRP composite. To determine the extent of the variability found in a typical set of witness panels, eight sets of FRP tensile testing data were obtained from several concrete strengthening projects (Kliger, personal communication, 2003). These projects were conducted in the United States between November 2001 and November 2003. All of the witness panels consisted of unidirectional carbon fiber reinforcement weighing approximately $4.27 \times 10^{-4}$ psi ($0.30 \, \text{kg/m}^2$) and were impregnated with an organic epoxy resin. However, slight variations in fabric style existed between sets.

Tensile data sets consisting of 143 tensile specimens are shown in Tables 4.1 to 4.8. The data provided for each set of tensile specimens were limited to specimen width, number of plies (layers), and maximum load. Using these values, the strength of the composite per unit length per ply can be calculated.

In addition to the raw data, the common summary statistics are presented at the bottom of each data set. These statistics include averages, standard

Table 4.1 Data set #1

| Specimen | # Plies | Width | | Maximum load | | Comp. strength per ply | |
|---|---|---|---|---|---|---|---|
| | | *(in.)* | *(cm)* | *(lb)* | *(kN)* | *(lb/in./ply)* | *(kN/cm/ply)* |
| 1 | 2 | 0.659 | 1.674 | 6027 | 26.81 | 4573 | 8.01 |
| 2 | 2 | 0.659 | 1.674 | 5369 | 23.88 | 4074 | 7.13 |
| 3 | 2 | 0.659 | 1.674 | 5777 | 25.70 | 4383 | 7.68 |
| 4 | 2 | 0.613 | 1.556 | 5683 | 25.28 | 4639 | 8.12 |
| 5 | 2 | 0.659 | 1.674 | 5133 | 22.83 | 3895 | 6.82 |
| 6 | 2 | 0.659 | 1.674 | 5885 | 26.18 | 4465 | 7.82 |
| 7 | 2 | 0.659 | 1.674 | 5815 | 25.87 | 4412 | 7.73 |
| 8 | 2 | 0.659 | 1.674 | 5842 | 25.99 | 4432 | 7.76 |
| 9 | 2 | 0.659 | 1.674 | 6400 | 28.47 | 4856 | 8.50 |
| 10 | 2 | 0.659 | 1.674 | 6042 | 26.87 | 4584 | 8.03 |
| 11 | 2 | 0.659 | 1.674 | 6586 | 29.29 | 4997 | 8.75 |
| 12 | 2 | 0.659 | 1.674 | 5752 | 25.58 | 4364 | 7.64 |
| Average = | | 0.655 | 1.664 | – | – | 4473 | 7.83 |
| Standard Dev. = | | 0.01342 | 0.03410 | – | – | 300 | 0.525 |
| C.O.V. = | | 2.0% | | – | | 6.7% | |

*Table 4.2* Data set #2

| Specimen | # Plies | Width | | Maximum load | | Comp. strength per ply | |
|---|---|---|---|---|---|---|---|
| | | (in.) | (cm) | (lb) | (kN) | (lb/in./ply) | (kN/cm/ply) |
| 1 | 1 | 0.638 | 1.620 | 2602 | 11.57 | 4080 | 7.14 |
| 2 | 1 | 0.628 | 1.595 | 2584 | 11.49 | 4115 | 7.21 |
| 3 | 2 | 0.638 | 1.620 | 5300 | 23.57 | 4155 | 7.28 |
| 4 | 2 | 0.634 | 1.610 | 5221 | 23.22 | 4118 | 7.21 |
| 5 | 2 | 0.634 | 1.610 | 5149 | 22.90 | 4062 | 7.11 |
| Average = | | 0.634 | 1.611 | – | – | 4106 | 7.19 |
| Standard Dev. = | | 0.00403 | 0.01025 | – | – | 36 | 0.064 |
| C.O.V. = | | 0.6% | | – | | 0.9% | |

*Table 4.3* Data set #3

| Specimen | # Plies | Width | | Maximum load | | Comp. strength per ply | |
|---|---|---|---|---|---|---|---|
| | | (in.) | (cm) | (lb) | (kN) | (lb/in./ply) | (kN/cm/ply) |
| 1 | 1 | 0.663 | 1.685 | 2418 | 10.76 | 3645 | 6.38 |
| 2 | 1 | 0.654 | 1.660 | 2420 | 10.76 | 3703 | 6.48 |
| 3 | 1 | 0.654 | 1.660 | 2938 | 13.07 | 4495 | 7.87 |
| 4 | 1 | 0.654 | 1.660 | 2627 | 11.68 | 4020 | 7.04 |
| 5 | 1 | 0.654 | 1.660 | 2262 | 10.06 | 3461 | 6.06 |
| 6 | 1 | 0.661 | 1.680 | 2574 | 11.45 | 3892 | 6.81 |
| Average = | | 0.656 | 1.668 | – | – | 3869 | 6.78 |
| Standard Dev. = | | 0.00462 | 0.01173 | – | – | 363 | 0.636 |
| C.O.V. = | | 0.7% | | – | | 9.4% | |

*Table 4.4* Data set #4

| Specimen | # Plies | Width | | Maximum load | | Comp. strength per ply | |
|---|---|---|---|---|---|---|---|
| | | (in.) | (cm) | (lb) | (kN) | (lb/in./ply) | (kN/cm/ply) |
| 1 | 1 | 0.535 | 1.360 | 2326 | 10.35 | 4344 | 7.61 |
| 2 | 1 | 0.535 | 1.360 | 2015 | 8.96 | 3763 | 6.59 |
| 3 | 1 | 0.634 | 1.610 | 2420 | 10.76 | 3818 | 6.69 |
| 4 | 1 | 0.575 | 1.460 | 1925 | 8.56 | 3349 | 5.86 |
| 5 | 1 | 0.535 | 1.360 | 2115 | 9.41 | 3950 | 6.92 |
| 6 | 1 | 0.535 | 1.360 | 2368 | 10.53 | 4423 | 7.74 |
| 7 | 1 | 0.531 | 1.350 | 2092 | 9.31 | 3936 | 6.89 |
| 8 | 1 | 0.535 | 1.360 | 2109 | 9.38 | 3939 | 6.90 |
| 9 | 1 | 0.535 | 1.360 | 2259 | 10.05 | 4219 | 7.39 |
| 10 | 1 | 0.634 | 1.610 | 2565 | 11.41 | 4047 | 7.09 |
| 11 | 1 | 0.634 | 1.610 | 2556 | 11.37 | 4032 | 7.06 |
| 12 | 1 | 0.528 | 1.340 | 1752 | 7.79 | 3321 | 5.82 |
| 13 | 1 | 0.528 | 1.340 | 2334 | 10.38 | 4424 | 7.75 |
| 14 | 1 | 0.520 | 1.320 | 2279 | 10.14 | 4385 | 7.68 |
| 15 | 1 | 0.520 | 1.320 | 1848 | 8.22 | 3556 | 6.23 |

Table 4.4 (Continued)

| Specimen | # Plies | Width | | Maximum load | | Comp. strength per ply | |
|---|---|---|---|---|---|---|---|
| | | (in.) | (cm) | (lb) | (kN) | (lb/in./ply) | (kN/cm/ply) |
| 16 | 1 | 0.520 | 1.320 | 2226 | 9.90 | 4283 | 7.50 |
| 17 | 1 | 0.528 | 1.340 | 1915 | 8.52 | 3630 | 6.36 |
| 18 | 1 | 0.516 | 1.310 | 1668 | 7.42 | 3234 | 5.66 |
| 19 | 1 | 0.516 | 1.310 | 2119 | 9.43 | 4109 | 7.19 |
| 20 | 1 | 0.531 | 1.350 | 1746 | 7.77 | 3285 | 5.75 |
| 21 | 1 | 0.524 | 1.330 | 2164 | 9.63 | 4133 | 7.24 |
| 22 | 1 | 0.642 | 1.630 | 2641 | 11.75 | 4115 | 7.21 |
| 23 | 1 | 0.618 | 1.570 | 2694 | 11.98 | 4358 | 7.63 |
| 24 | 1 | 0.646 | 1.640 | 2690 | 11.97 | 4166 | 7.30 |
| 25 | 1 | 0.634 | 1.610 | 2325 | 10.34 | 3668 | 6.42 |
| 26 | 1 | 0.520 | 1.320 | 1502 | 6.68 | 2890 | 5.06 |
| 27 | 1 | 0.520 | 1.320 | 2116 | 9.41 | 4072 | 7.13 |
| 28 | 1 | 0.524 | 1.330 | 1772 | 7.88 | 3384 | 5.93 |
| 29 | 1 | 0.520 | 1.320 | 2074 | 9.23 | 3991 | 6.99 |
| 30 | 1 | 0.634 | 1.610 | 2556 | 11.37 | 4032 | 7.06 |
| 31 | 1 | 0.634 | 1.610 | 2560 | 11.39 | 4039 | 7.07 |
| 32 | 1 | 0.524 | 1.330 | 1742 | 7.75 | 3327 | 5.83 |
| 33 | 1 | 0.520 | 1.320 | 2115 | 9.41 | 4070 | 7.13 |
| 34 | 1 | 0.634 | 1.610 | 2513 | 11.18 | 3965 | 6.94 |
| 35 | 1 | 0.634 | 1.610 | 2600 | 11.56 | 4102 | 7.18 |
| Average = | | 0.561 | 1.426 | – | – | 3896 | 6.82 |
| Standard Dev. = | | 0.05112 | 0.12985 | – | – | 391 | 0.684 |
| C.O.V. = | | 9.1% | | – | | 10.0% | |

deviations, and coefficients of variation. The average value or mean, $\overline{x}$, of any set of data was simply the sum of the data divided by the number of data points,

$$\overline{x} = \frac{\sum_{i=1}^{n} x_i}{n} \tag{4.1}$$

where $x_i$ is each individual data value and $n$ is the total number of data points or observations within the set. For each set of data, the sample standard deviation, $s$, was computed using the following formula:

$$s = \sqrt{\frac{\sum_{i=1}^{n} (x_i - \overline{x})^2}{n - 1}} \tag{4.2}$$

To measure the reliability of a data set, a dimensionless parameter known as the coefficient of variation was used. This parameter expresses variation as a fraction of the mean and is defined as:

$$COV = \frac{s}{\overline{x}} \tag{4.3}$$

Table 4.5 Data set #5

| Specimen | # Plies | Width | | Maximum load | | Comp. strength per ply | |
|---|---|---|---|---|---|---|---|
| | | (in.) | (cm) | (lb) | (kN) | (lb/in./ply) | (kN/cm/ply) |
| 1 | 2 | 0.535 | 1.360 | 3650 | 16.24 | 3408 | 5.97 |
| 2 | 2 | 0.535 | 1.360 | 3490 | 15.52 | 3259 | 5.71 |
| 3 | 2 | 0.528 | 1.340 | 6074 | 27.02 | 5757 | 10.08 |
| 4 | 2 | 0.531 | 1.350 | 5962 | 26.52 | 5609 | 9.82 |
| 5 | 2 | 0.531 | 1.350 | 4256 | 18.93 | 4004 | 7.01 |
| 6 | 2 | 0.531 | 1.350 | 4450 | 19.79 | 4186 | 7.33 |
| 7 | 2 | 0.531 | 1.350 | 3473 | 15.45 | 3267 | 5.72 |
| 8 | 2 | 0.531 | 1.350 | 3186 | 14.17 | 2997 | 5.25 |
| 9 | 2 | 0.531 | 1.350 | 4540 | 20.19 | 4271 | 7.48 |
| 10 | 2 | 0.531 | 1.350 | 4159 | 18.50 | 3913 | 6.85 |
| 11 | 2 | 0.516 | 1.310 | 3859 | 17.16 | 3741 | 6.55 |
| 12 | 2 | 0.531 | 1.350 | 2899 | 12.89 | 2727 | 4.78 |
| 13 | 2 | 0.634 | 1.610 | 4746 | 21.11 | 3744 | 6.56 |
| 14 | 2 | 0.634 | 1.610 | 5431 | 24.16 | 4284 | 7.50 |
| 15 | 2 | 0.634 | 1.610 | 7723 | 34.35 | 6092 | 10.67 |
| 16 | 2 | 0.634 | 1.610 | 6163 | 27.41 | 4861 | 8.51 |
| 17 | 2 | 0.630 | 1.600 | 5218 | 23.21 | 4142 | 7.25 |
| 18 | 2 | 0.634 | 1.610 | 5414 | 24.08 | 4271 | 7.48 |
| 19 | 2 | 0.630 | 1.600 | 6084 | 27.06 | 4829 | 8.46 |
| 20 | 2 | 0.634 | 1.610 | 7818 | 34.77 | 6167 | 10.80 |
| 21 | 2 | 0.622 | 1.580 | 6096 | 27.12 | 4900 | 8.58 |
| 22 | 2 | 0.634 | 1.610 | 6363 | 28.30 | 5019 | 8.79 |
| 23 | 2 | 0.634 | 1.610 | 3994 | 17.77 | 3151 | 5.52 |
| 24 | 2 | 0.630 | 1.600 | 4471 | 19.89 | 3549 | 6.21 |
| 25 | 3 | 0.642 | 1.630 | 6627 | 29.48 | 3442 | 6.03 |
| 26 | 3 | 0.642 | 1.630 | 6869 | 30.55 | 3568 | 6.25 |
| Average = | | 0.586 | 1.488 | – | – | 4198 | 7.35 |
| Standard Dev. = | | 0.05248 | 0.13330 | – | – | 957 | 1.676 |
| C.O.V. = | | 9.0% | | – | | 22.8% | |

Table 4.6 Data set #6

| Specimen | # Plies | Width | | Maximum load | | Comp. strength per ply | |
|---|---|---|---|---|---|---|---|
| | | (in.) | (cm) | (lb) | (kN) | (lb/in./ply) | (kN/cm/ply) |
| 1 | 2 | 0.661 | 1.680 | 4871 | 21.67 | 3682 | 6.45 |
| 2 | 2 | 0.661 | 1.680 | 4188 | 18.63 | 3166 | 5.54 |
| 3 | 2 | 0.659 | 1.674 | 6027 | 26.81 | 4573 | 8.01 |
| 4 | 2 | 0.659 | 1.674 | 5369 | 23.88 | 4074 | 7.13 |
| 5 | 2 | 0.659 | 1.674 | 5777 | 25.70 | 4383 | 7.68 |
| 6 | 2 | 0.613 | 1.556 | 5683 | 25.28 | 4639 | 8.12 |
| 7 | 2 | 0.659 | 1.674 | 5133 | 22.83 | 3895 | 6.82 |
| 8 | 2 | 0.659 | 1.674 | 5885 | 26.18 | 4465 | 7.82 |
| 9 | 2 | 0.659 | 1.674 | 5815 | 25.87 | 4412 | 7.73 |

*Table 4.6* (Continued)

| Specimen | # Plies | Width | | Maximum load | | Comp. strength per ply | |
|---|---|---|---|---|---|---|---|
| | | *(in.)* | *(cm)* | *(lb)* | *(kN)* | *(lb/in./ply)* | *(kN/cm/ply)* |
| 10 | 2 | 0.659 | 1.674 | 5842 | 25.99 | 4432 | 7.76 |
| 11 | 2 | 0.659 | 1.674 | 6400 | 28.47 | 4856 | 8.50 |
| 12 | 2 | 0.659 | 1.674 | 6042 | 26.87 | 4584 | 8.03 |
| 13 | 2 | 0.659 | 1.674 | 6586 | 29.29 | 4997 | 8.75 |
| 14 | 2 | 0.659 | 1.674 | 5752 | 25.58 | 4364 | 7.64 |
| Average = | | 0.656 | 1.666 | – | – | 4323 | 7.57 |
| Standard Dev. = | | 0.01256 | 0.03189 | – | – | 481 | 0.843 |
| C.O.V. = | | 1.9% | | – | | 11.1% | |

*Table 4.7* Data set #7

| Specimen | # Plies | Width | | Maximum load | | Comp. strength per ply | |
|---|---|---|---|---|---|---|---|
| | | *(in.)* | *(cm)* | *(lb)* | *(kN)* | *(lb/in./ply)* | *(kN/cm/ply)* |
| 1 | 1 | 0.638 | 1.620 | 2602 | 11.57 | 4080 | 7.14 |
| 2 | 1 | 0.642 | 1.630 | 2064 | 9.18 | 3216 | 5.63 |
| 3 | 2 | 0.638 | 1.620 | 5300 | 23.57 | 4155 | 7.28 |
| 4 | 2 | 0.634 | 1.610 | 5221 | 23.22 | 4118 | 7.21 |
| Average = | | 0.638 | 1.620 | – | – | 3892 | 6.82 |
| Standard Dev. = | | 0.00321 | 0.00816 | – | – | 452 | 0.791 |
| C.O.V. = | | 0.5% | | – | | 11.6% | |

*Table 4.8* Data set #8

| Specimen | # Plies | Width | | Maximum load | | Comp. strength per ply | |
|---|---|---|---|---|---|---|---|
| | | *(in.)* | *(cm)* | *(lb)* | *(kN)* | *(lb/in./ply)* | *(kN/cm/ply)* |
| 1 | 3 | 0.640 | 1.626 | 5414 | 24.08 | 2820 | 4.94 |
| 2 | 3 | 0.654 | 1.661 | 5755 | 25.60 | 2933 | 5.14 |
| 3 | 3 | 0.652 | 1.656 | 5962 | 26.52 | 3048 | 5.34 |
| 4 | 3 | 0.651 | 1.654 | 5259 | 23.39 | 2693 | 4.72 |
| 5 | 3 | 0.642 | 1.631 | 5094 | 22.66 | 2645 | 4.63 |
| 6 | 3 | 0.653 | 1.659 | 6490 | 28.87 | 3313 | 5.80 |
| 7 | 3 | 0.653 | 1.659 | 5296 | 23.56 | 2703 | 4.73 |
| 8 | 3 | 0.656 | 1.666 | 6081 | 27.05 | 3090 | 5.41 |
| 9 | 3 | 0.650 | 1.651 | 6055 | 26.93 | 3105 | 5.44 |
| 10 | 3 | 0.653 | 1.659 | 6146 | 27.34 | 3137 | 5.49 |
| 11 | 3 | 0.652 | 1.656 | 5825 | 25.91 | 2978 | 5.22 |
| 12 | 3 | 0.653 | 1.659 | 5763 | 25.63 | 2942 | 5.15 |
| 13 | 3 | 0.649 | 1.648 | 5887 | 26.19 | 3024 | 5.29 |
| 14 | 3 | 0.651 | 1.654 | 5488 | 24.41 | 2810 | 4.92 |

| | | | | | | | |
|---|---|---|---|---|---|---|---|
| 15 | 3 | 0.653 | 1.659 | 5853 | 26.03 | 2988 | 5.23 |
| 16 | 3 | 0.653 | 1.659 | 6100 | 27.13 | 3114 | 5.45 |
| 17 | 3 | 0.652 | 1.656 | 6529 | 29.04 | 3338 | 5.85 |
| 18 | 3 | 0.651 | 1.654 | 6001 | 26.69 | 3073 | 5.38 |
| 19 | 3 | 0.651 | 1.654 | 5831 | 25.94 | 2986 | 5.23 |
| 20 | 3 | 0.640 | 1.626 | 5916 | 26.31 | 3081 | 5.40 |
| 21 | 3 | 0.636 | 1.615 | 6009 | 26.73 | 3149 | 5.52 |
| 22 | 3 | 0.644 | 1.636 | 6225 | 27.69 | 3222 | 5.64 |
| 23 | 3 | 0.644 | 1.636 | 6403 | 28.48 | 3314 | 5.80 |
| 24 | 3 | 0.652 | 1.656 | 6180 | 27.49 | 3160 | 5.53 |
| 25 | 3 | 0.651 | 1.654 | 6666 | 29.65 | 3413 | 5.98 |
| 26 | 3 | 0.653 | 1.659 | 6444 | 28.66 | 3289 | 5.76 |
| 27 | 3 | 0.651 | 1.654 | 5927 | 26.36 | 3035 | 5.31 |
| 28 | 3 | 0.646 | 1.641 | 6809 | 30.29 | 3513 | 6.15 |
| 29 | 3 | 0.654 | 1.661 | 6398 | 28.46 | 3261 | 5.71 |
| 30 | 3 | 0.652 | 1.656 | 6357 | 28.28 | 3250 | 5.69 |
| 31 | 3 | 0.652 | 1.656 | 6707 | 29.83 | 3429 | 6.00 |
| 32 | 3 | 0.652 | 1.656 | 6726 | 29.92 | 3439 | 6.02 |
| 33 | 3 | 0.652 | 1.656 | 6822 | 30.34 | 3488 | 6.11 |
| 34 | 3 | 0.654 | 1.661 | 5855 | 26.04 | 2984 | 5.23 |
| 35 | 3 | 0.655 | 1.664 | 5361 | 23.85 | 2728 | 4.78 |
| 36 | 3 | 0.647 | 1.643 | 5393 | 23.99 | 2778 | 4.87 |
| 37 | 3 | 0.655 | 1.664 | 5742 | 25.54 | 2922 | 5.12 |
| 38 | 3 | 0.651 | 1.654 | 6031 | 26.83 | 3088 | 5.41 |
| 39 | 3 | 0.655 | 1.664 | 5406 | 24.05 | 2751 | 4.82 |
| 40 | 3 | 0.652 | 1.656 | 6308 | 28.06 | 3225 | 5.65 |
| 41 | 3 | 0.653 | 1.659 | 6314 | 28.08 | 3223 | 5.64 |
| Average = | | 0.650 | 1.652 | – | – | 3085 | 5.40 |
| Standard Dev. = | | 0.00453 | 0.01152 | – | – | 231 | 0.404 |
| C.O.V. = | | | 0.7% | | – | | 7.5% |

The lower the coefficient of variation, the more reliable the data can be considered (Montgomery and Runger, 1994).

### 4.8.4 *Analysis and discussion*

Table 4.9 presents a summary of the basic statistics for each of the eight sets of data. To obtain an overall measure of variability for strength and width, an average coefficient of variation is desired. However, a simple average value is not practical since each set of data contains a different number of observations. Data sets containing many specimens should have more influence on the average than data sets containing only a few specimens. Therefore, a weighted average was computed using the following formula:

$$\overline{\overline{x}} = \frac{\sum_{i=1}^{n} (\overline{x}_i n_i)}{\sum_{i=1}^{n} n_i} \tag{4.4}$$

Table 4.9 Summary statistics for data sets

| Data Set # | # Specimens, n | Averages | | | | Standard deviations | | | | C.O.V. | |
|---|---|---|---|---|---|---|---|---|---|---|---|
| | | Width | | Strength | | Width | | Strength | | Width | Comp. strength per ply |
| | | (cm) | (in.) | (kN/cm/ply) | (lb/in./ply) | (cm) | (in.) | (kN/cm/ply) | (lb/in./ply) | | |
| 1 | 12 | 1.6640 | 0.6551 | 7.83 | 4473 | 0.0341 | 0.0134 | 0.525 | 300 | 2.0% | 6.7% |
| 2 | 5 | 1.6110 | 0.6343 | 7.19 | 4106 | 0.0102 | 0.0040 | 0.064 | 36 | 0.6% | 0.9% |
| 3 | 6 | 1.6675 | 0.6565 | 6.78 | 3869 | 0.0117 | 0.0046 | 0.636 | 363 | 0.7% | 9.4% |
| 4 | 35 | 1.4260 | 0.5614 | 6.82 | 3896 | 0.1298 | 0.0511 | 0.684 | 391 | 9.1% | 10.0% |
| 5 | 26 | 1.4881 | 0.5859 | 7.35 | 4198 | 0.1333 | 0.0525 | 1.676 | 957 | 9.0% | 22.8% |
| 6 | 14 | 1.6663 | 0.6560 | 7.57 | 4323 | 0.0319 | 0.0126 | 0.843 | 481 | 1.9% | 11.1% |
| 7 | 4 | 1.6200 | 0.6378 | 6.82 | 3892 | 0.0082 | 0.0032 | 0.791 | 452 | 0.5% | 11.6% |
| 8 | 41 | 1.6522 | 0.6505 | 5.40 | 3085 | 0.0115 | 0.0045 | 0.404 | 231 | 0.7% | 7.5% |

*Table 4.10* Calculations performed for weighted averages

| Data Set # | # Specimens, n | Comp. strength per ply | | Width | |
|---|---|---|---|---|---|
| | | C.O.V | C.O.V*n | C.O.V | C.O.V*n |
| 1 | 12 | 0.0671 | 0.81 | 0.0205 | 0.246 |
| 2 | 5 | 0.0089 | 0.04 | 0.0064 | 0.032 |
| 3 | 6 | 0.0939 | 0.56 | 0.0070 | 0.042 |
| 4 | 35 | 0.1003 | 3.51 | 0.0911 | 3.187 |
| 5 | 26 | 0.2280 | 5.93 | 0.0896 | 2.329 |
| 6 | 14 | 0.1113 | 1.56 | 0.0191 | 0.268 |
| 7 | 4 | 0.1161 | 0.46 | 0.0050 | 0.020 |
| 8 | 41 | 0.0747 | 3.06 | 0.0070 | 0.286 |
| Sums, Σ | 143 | | 15.94 | | 6.41 |
| Weighted averages | | 15.94/143 = | 11.1% | 6.41/143 = | 4.5% |

where $\overline{\overline{x}}$ is the weighted average. Substituting the strength data into Equation (4.4) leads to the following:

$$\overline{\overline{x}} = \frac{\begin{aligned}(0.067 \times 12) + (0.009 \times 5) + (0.094 \times 6) + (0.10 \times 35) + \\ (0.228 \times 26) + (0.111 \times 14) + (0.116 \times 4) + (0.075 \times 41)\end{aligned}}{(12 + 5 + 6 + 35 + 26 + 14 + 4 + 41)}$$

$$= 0.111 = 11.1\% \tag{4.5}$$

The average coefficient of variation for composite strength was 11.1%. Performing a similar calculation for the width values, it was determined that the average coefficient of variation was approximately 4.5%. Further details of these computations are tabulated in Tables 4.9 and 4.10.

## 4.9 Case study

This section presents tensile results from E-glass witness panels manufactured during an FRP rehabilitation of a bridge overpass in Rhode Island. In this project, a thin E-glass fabric was impregnated with an organic resin and placed on the pier cap beams of the Silver Spring Cove Bridge. Vacuum bagging was utilized to achieve a good bond between the composite and the concrete structure. The tensile results that follow were part of the quality control measures used during the implementation of the composite jacket.

### 4.9.1 Specimen preparation

The E-glass witness panels were prepared on site using the same materials that were used for the composite jacket. The composite consisted of two plies of fiberglass fabric impregnated with the polymer resin, Sikadur® Hex 300.

*Table 4.11* Details of tabbing epoxy

| | |
|---|---|
| Form: | Two-part paste adhesive |
| Color: | Pale blue |
| Viscosity: | 250,000 cps |
| Volumetric mixing ratio A:B | 1:1 |
| Setting time (Pot Life): | 30 min at 25°C (77°F) |
| Service temperature range: | −10°C (−50°F) to +66°C (+150°F) |
| Time to full cure at 25°C (77°F): | 24 h |
| Tensile shear strength at 24°C (75°F): | 22.75 MPa (3300 psi) |

The wet fiberglass was placed in between two plies of nonporous, nonstick, Teflon® fabric. After 72 h of curing at the job site, the cured witness panel was removed from the Teflon. Straight-sided coupons measuring approximately 0.65 in. (1.65 cm) wide, 12 in. (30.5 cm) long, and 0.1 in. (0.25 cm) thick were cut from the witness panel using a diamond blade wet saw.

Testing the fiberglass coupons in tension required that the ends of the specimens be reinforced with aluminum tabs to prevent grip failures. When straight-sided specimens are used, stress concentrations from gripping will likely cause failure. Therefore, many composite specimens require tabs to be bonded to both sides at the ends. These tabs distribute the gripping stresses and prevent failure caused by the jaws of the grip damaging the specimen's surface (Instron, 2003).

To ensure a good bond between the aluminum tabs and the specimen, special care was taken to prepare the bonding surfaces of both the tab and the composite. First, sandpaper (grit 150) was used to roughen the ends of each specimen where tabbing would take place. All sanded surfaces were then thoroughly washed with a generous amount of acetone to remove any debris, oil, and so on. The aluminum tabs measured 0.65 in. (1.65 cm) wide, 2.4 in. (6.1 cm) long, and 0.08 in. (0.20 cm) thick and were prepared in a similar manner. The surface of each aluminum tab was abraded using a wire wheel, followed by an acetone wash to cleanse the surface of impurities.

To bond the aluminum tabs to the fiberglass coupon, a room-temperature-curing epoxy was used. Some of the essential features and mechanical properties of the two-part commercially available epoxy are listed in Table 4.11. Equal volumes of parts A and B were mixed together and a generous amount of epoxy was applied to both the tab and the specimen. The tabs were then secured onto the coupon using butterfly clips and were allowed to cure in the laboratory for 72 h at approximately 75°F (24°C).

The presence of any impurities can severely impair the bonding between the aluminum tabs and the fiberglass composite. Therefore, throughout the entire tabbing process, the specimens were always handled from the gage length, *not* by the specimen ends. This avoids contaminating the bonding surfaces with grease, oil, and other debris. Since even a fingerprint can compromise the bond, latex gloves were always worn.

## 4.9.2 Test method

Each tensile specimen was tested to determine the in-plane tensile properties. ASTM D3039, "Tensile Properties of Polymer Matrix Composite Materials" was used as a guideline (American Society for Testing and Materials, 2006). The tensile samples were mounted in self-aligning grips of an MTS Sintech 10/GL mechanical testing machine and continuously loaded in tension while recording load to the nearest 0.1 lb (0.5 N). The samples were tested using a constant head displacement of 0.05 in./min (1.27 mm/min). Twenty specimens were tested and, on average, each test was completed in 5 min. Using digital calipers, the width and thickness of each specimen were measured at the failure location to within 0.0005 in. (0.013 mm). In addition, the failure mode of each specimen was recorded. The laboratory temperature during the tests was approximately 72 °F (22 °C).

## 4.9.3 Test results

The maximum failure load was recorded using the MTS system and the corresponding TestWorks® software. The specimen geometry, maximum loads, and failure codes are presented in Table 4.12 while Table 4.13 includes the strength computations. Table 4.14 is a summary of the failure modes exhibited by all of the specimens.

### 4.9.3.1 Strength

Since both the matrix and the fibers contribute to the strength, equivalent fabric strength was computed using the subsequent procedure. The following symbols were utilized:

$V_f$ = volume fraction of fiber
$V_m$ = volume fraction of matrix (resin)
$V_c$ = volume fraction of composite
$\gamma_f$ = density of fibers
$\gamma_m$ = density of matrix
$\gamma_c$ = density of composite

and the Law of Mixtures:

$$V_c\gamma_c = V_f\gamma_f + V_m\gamma_m \tag{4.6}$$

The density of the composite specimen, $\gamma_c$, as well as that of the dry fabric, $\gamma_f$, was determined using the density by displacement method (American Society for Testing and Materials, 2000). Since the density of the matrix was

*Table 4.12* Specimen geometry and test results for Rhode Island bridge witness panels

| Specimen ID | Thickness, t | | Width, w | | Maximum load, $P_{max}$ | | Failure code |
|---|---|---|---|---|---|---|---|
| | *(mm)* | *(in.)* | *(cm)* | *(in.)* | *(N)* | *(lb)* | |
| 1 | 0.2667 | 0.0105 | 1.6713 | 0.6580 | 1001 | 225.0 | LWT |
| 2 | 0.2667 | 0.0105 | 1.6675 | 0.6565 | 738 | 166.0 | AWT |
| 3 | 0.2159 | 0.0085 | 1.6485 | 0.6490 | 707 | 159.0 | LWB |
| 4 | 0.2413 | 0.0095 | 1.6485 | 0.6490 | 752 | 169.0 | AWB |
| 5 | 0.2413 | 0.0095 | 1.6129 | 0.6350 | 796 | 179.0 | LGM |
| 6 | 0.2794 | 0.0110 | 1.6459 | 0.6480 | 1013 | 227.8 | LWT |
| 7 | 0.2794 | 0.0110 | 1.6586 | 0.6530 | 990 | 222.5 | LGM |
| 8 | 0.2540 | 0.0100 | 1.6447 | 0.6475 | 935 | 210.3 | LGM |
| 9 | 0.2794 | 0.0110 | 1.6675 | 0.6565 | 1056 | 237.4 | LGM |
| 10 | 0.2413 | 0.0095 | 1.6612 | 0.6540 | 1040 | 233.8 | AWB |
| 11 | 0.2286 | 0.0090 | 1.6828 | 0.6625 | 1049 | 235.9 | LGM |
| 12 | 0.2540 | 0.0100 | 1.6421 | 0.6465 | 1023 | 230.0 | LGM |
| 13 | 0.2540 | 0.0100 | 1.6624 | 0.6545 | 1048 | 235.5 | LGM |
| 14 | 0.2540 | 0.0100 | 1.6497 | 0.6495 | 1092 | 245.4 | LWT |
| 15 | 0.2413 | 0.0095 | 1.6599 | 0.6535 | 997 | 224.1 | LAB |
| 16 | 0.2667 | 0.0105 | 1.6370 | 0.6445 | 958 | 215.4 | LAB |
| 17 | 0.2540 | 0.0100 | 1.5875 | 0.6250 | 758 | 170.5 | AAT |
| 18 | 0.2413 | 0.0095 | 1.6650 | 0.6555 | 983 | 221.0 | LGM |
| 19 | 0.2286 | 0.0090 | 1.6523 | 0.6505 | 971 | 218.3 | LGM |
| 20 | 0.2540 | 0.0100 | 1.6586 | 0.6530 | 1014 | 228.0 | LWB |
| Average | 0.2521 | 0.0099 | 1.6512 | 0.6501 | 946 | 213 | – |
| Std. Dev. | 0.0176 | 0.0007 | 0.0211 | 0.0083 | 122.1 | 27.5 | – |

determined to be $\gamma_m = 0.04\,lb/in.^3$ (1.13 g/cm$^3$), the volume fraction of the matrix was computed using the equation:

$$V_m = \frac{V_c \gamma_c - V_f \gamma_f}{\gamma_m} \tag{4.7}$$

Since the fabric and the matrix are assumed the only components of the composite, the volume fraction of fiber can be written as:

$$V_f = 1 - V_m \tag{4.8}$$

Substituting Equation (4.8) into Equation (4.7) yields:

$$V_m = \frac{\gamma_c - \gamma_f}{\gamma_m - \gamma_f} \tag{4.9}$$

Using this equation, it was determined that the average volume fraction of matrix, $V_m$, was approximately 24.6%. Substituting this value into Equation (4.8) led to a volume fraction of fiber of about 75.4%.

*Table 4.13* Tensile strength results for Rhode Island bridge project witness panels

| Specimen ID | Composite strength, $S_c$ | | | | Composite strength per ply | | Stress in fabric, $S_f$ | |
|---|---|---|---|---|---|---|---|---|
| | (MPa) | (ksi) | (N/cm) | (lb/in.) | (N/cm/ply) | (lb/in./ply) | (MPa) | (ksi) |
| 1 | 225 | 32.56 | 599 | 342 | 299 | 171 | 596 | 86.40 |
| 2 | 166 | 24.08 | 443 | 253 | 221 | 126 | 441 | 63.89 |
| 3 | 199 | 28.82 | 429 | 245 | 215 | 122 | 527 | 76.47 |
| 4 | 189 | 27.40 | 456 | 260 | 228 | 130 | 502 | 72.72 |
| 5 | 205 | 29.66 | 494 | 282 | 247 | 141 | 543 | 78.72 |
| 6 | 220 | 31.95 | 616 | 352 | 308 | 176 | 585 | 84.79 |
| 7 | 214 | 30.97 | 597 | 341 | 298 | 170 | 567 | 82.18 |
| 8 | 224 | 32.47 | 569 | 325 | 284 | 162 | 594 | 86.17 |
| 9 | 227 | 32.87 | 633 | 362 | 317 | 181 | 601 | 87.22 |
| 10 | 259 | 37.62 | 626 | 357 | 313 | 179 | 689 | 99.84 |
| 11 | 273 | 39.55 | 624 | 356 | 312 | 178 | 724 | 104.97 |
| 12 | 245 | 35.57 | 623 | 356 | 312 | 178 | 651 | 94.39 |
| 13 | 248 | 35.97 | 630 | 360 | 315 | 180 | 658 | 95.46 |
| 14 | 261 | 37.77 | 662 | 378 | 331 | 189 | 691 | 100.24 |
| 15 | 249 | 36.09 | 601 | 343 | 300 | 171 | 660 | 95.77 |
| 16 | 219 | 31.82 | 585 | 334 | 293 | 167 | 582 | 84.45 |
| 17 | 188 | 27.27 | 478 | 273 | 239 | 136 | 499 | 72.38 |
| 18 | 245 | 35.48 | 590 | 337 | 295 | 169 | 649 | 94.16 |
| 19 | 257 | 37.28 | 588 | 336 | 294 | 168 | 682 | 98.93 |
| 20 | 241 | 34.91 | 611 | 349 | 306 | 175 | 639 | 92.63 |
| Maximum | 273 | 39.55 | 662 | 378 | 331 | 189 | 724 | 105 |
| Minimum | 166 | 24.08 | 429 | 245 | 215 | 122 | 441 | 64 |
| Average | 228 | 33.01 | 573 | 327 | 286 | 163 | 604 | 88 |
| Std. Dev. | 28.3 | 4.10 | 70.8 | 40.4 | 35.4 | 20.2 | 75.1 | 10.9 |
| C.O.V. | | 12.4% | | | | 12.4% | | 12.4% |

*Table 4.14* Specimen geometry and test results for Rhode Island bridge witness panels

| First character | | Second character | | Third character | |
|---|---|---|---|---|---|
| Failure type | | Failure area | | Failure location | |
| 80% | Lateral | 45% | Gage | 25% | Top |
| 20% | Angled | 40% | <1W from Tab | 45% | Middle |
| | | 15% | At grip/tab | 30% | Bottom |

The composite strength, $S_c$, is based upon the gross cross-sectional area of the specimen and is determined using:

$$S_c = \frac{P_{max}}{tw} \tag{4.10}$$

where $P_{max}$ is the maximum recorded load and $t$ and $w$ are the specimen thickness and width, respectively. It is common to express the composite strength per unit width. In this case, $S_c$ is simply divided by the specimen width. To express the strength as a function of width and number of plies, $S_c$ is divided by both the specimen width and the number of reinforcement layers (in this case, all specimens consisted of two plies). The strength of the composite can be used to determine the strength of the fabric, $S_f$, using the following equation:

$$S_f = \frac{S_c}{V_f} \frac{1}{V_{fL}}$$

(4.11)

where $V_{fL}$ is the percentage of fibers in the loading direction of the reinforcement layer (~100% for unidirectional fabrics, ~50% for woven fabrics). Since this bidirectional E-glass fabric provided 50% of the fibers in the loading direction, $V_{fL} = 50\%$. The composite strength and equivalent fabric stresses are presented in Table 4.13. In addition, the composite strength as a function of both width and number of plies is also presented.

### 4.9.3.2  Failure codes

Since the mode of failure of a tensile specimen cannot be quantified, the American Society for Testing and Materials devised a standard failure code system, as shown in Figure 4.7. The failure code for a particular specimen consists of three separate categories or characters, each denoted by a letter value. The first character represents the nature or *type* of failure (grip, delamination, lateral, longitudinal splitting, explosive, and so on) while the second characterizes the failure *area*, such as "at grip/tab," and "gage." The third character distinguishes the failure *location*, namely, the top, bottom, left, right, and so on. Although these failure codes cover most tensile failures, not all failure modes will conform to these standard representations. Nine of the most common failure modes are shown in Figure 4.7.

### 4.9.4  Discussion and observations

By observing the summary of failure modes in Table 4.14, it can be seen that a majority of the specimens (80%) exhibited a "lateral" type of failure, which is a break in the specimen perpendicular to the longitudinal axis. None of the specimens exhibited a "grip/tab" failure, indicating that the bonded tabs were effective in preventing grip failures. In terms of failure area, the predominant result was failure at the gage length, which is the ideal result. Almost half of the specimens (45%) exhibited a failure located in the middle of the specimen while the other failure locations were approximately evenly divided between the top and bottom, 25 and 30%, respectively.

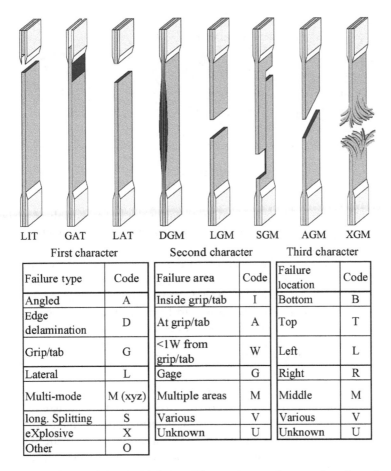

LIT  GAT  LAT  DGM  LGM  SGM  AGM  XGM

First character    Second character    Third character

| Failure type | Code | Failure area | Code | Failure location | Code |
|---|---|---|---|---|---|
| Angled | A | Inside grip/tab | I | Bottom | B |
| Edge delamination | D | At grip/tab | A | Top | T |
| Grip/tab | G | <1W from grip/tab | W | Left | L |
| Lateral | L | Gage | G | Right | R |
| Multi-mode | M (xyz) | Multiple areas | M | Middle | M |
| long. Splitting | S | Various | V | Various | V |
| eXplosive | X | Unknown | U | Unknown | U |
| Other | O | | | | |

*Figure 4.7* Failure codes for tensile specimens (Source: American Society for Testing and Materials, 2006).

The failure modes for two specimens are displayed in Figure 4.8. Since the failure mode is often an indicator of strength, it is expected that sample #11 which exhibited the ideal "LGM" failure mode was stronger than sample #17, which displayed an unusual "AAT" failure. This strength expectation can be verified by comparing the composite strength for the two specimens in Table 4.13. Specimen #11 achieved a tensile strength of 39.6 ksi (273 MPa) while specimen #17 obtained a strength of 27.3 ksi (188 MPa), which is only 69% of specimen #11.

From the results presented in Table 4.13, the average composite strength was 33 ksi (228 MPa) while the average maximum stress in the fiberglass fabric was approximately 87.6 ksi (604 MPa). The fabric stresses derived from the composite tension tests are lower than the actual strength of the fabric, 100 ksi (690 MPa), because some fibers can be damaged, snagged, and

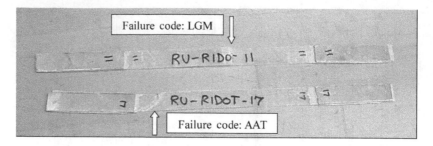

*Figure 4.8* Failure of two tensile specimens.

torn during the laminating process. The stress in the fiberglass fabric varied between 64 and 105 ksi (441 and 724 MPa) with a standard deviation of 10.9 ksi (75 MPa).

To statistically measure the variation of the specimen data, the coefficient of variation was calculated for composite strength and specimen width using Equation (4.3). As presented earlier in this chapter, the average coefficients of variation for composite strength per unit width per ply and specimen width in a typical field application were approximately 11.1 and 4.5%, respectively. For this project, the coefficients of variation for strength and width were 12.4 and 1.3%, respectively, which indicates that the amount of variation was about the same as in other bridge projects.

Minimal variations are acceptable and can be attributed to a variety of factors. Most commonly, the manufacturing process used to impregnate fiber with resin leads to varying degrees of imperfections and air inclusions (SP Systems, 2001). In addition, during the hand-impregnation of the witness panels, a small fraction of the fibers are damaged and/or destroyed. The geometry of the fibers in a composite is also important since fibers have their highest mechanical properties along their lengths. This leads to the highly anisotropic properties of composites where the mechanical behavior of the composite is likely to be very different when tested in different directions. Therefore, if the fibers in the tension specimen are even slightly off-axis, the strength measurements will be significantly different from a specimen with fibers oriented perfectly in the longitudinal direction.

One of the most important factors affecting the mechanical properties of the composite is the ratio of fiber to resin in the composite. Since the mechanical properties of fibers are usually much greater than those of polymers, the higher the volume fraction of fiber, the higher will be the mechanical properties of the resultant composite. In practice, however, there are certain limits to this since the fibers need to be fully saturated with resin to become effective. Typically, with a common hand lay-up process, a limit for the volume fraction of fiber is approximately 30–40%. With the higher quality, more sophisticated and precise manufacturing processes used in the aerospace industry, volume fractions of fiber approaching 70% can be successfully obtained (SP Systems, 2001). The composite jacket placed on the Silver Spring Cove Bridge had a volume fraction of fiber equal to approximately 75%, which is characteristic of aerospace quality and a superior manufacturing technique.

# References

ACI Committee 440 (1996) *State-of-the-Art Report on Fiber Reinforced Plastic (FRP) Reinforcement for Concrete Structures (ACI 440 R-96)*. Farmington Hills, MI: American Concrete Institute, 68 pp.

ACI Committee 440 (2002) *Guide for the Design and Construction of Externally Bonded FRP Systems for Strengthening Concrete Structures (ACI 440.2R-02)*. Detroit, MI: American Concrete Institute, 45 pp.

ACI Committee 318 (2005) *Building Code Requirements for Structural Concrete and Commentary (ACI 318-05)*. Farmington Hills, MI: American Concrete Institute, 430 pp.

American Society for Testing and Materials (2000) ASTM Test Method D792: 2001: *Standard Test Methods for Density and Specific Gravity (Relative Density) of Plastics by Displacement*. Pennsylvania: ASTM.

American Society for Testing and Materials (2006) ASTM Test Method D3039: 2006: *Standard Test Method for Tensile Properties of Polymer Matrix Composite Materials*. Pennsylvania: ASTM.

Arduini, M. and Nanni, A. (1997) Behavior of precracked RC beams strengthened with carbon FRP sheets. *Journal of Composites for Construction*, 1(2), pp. 63–70.

Chajes, M., Januska, T., Mertz, D., Thomson, T., and Finch, W. (1995) Shear strengthening of reinforced concrete beams using externally applied composite fabrics. *ACI Structural Journal*, 92(3), pp. 295–303.

Edge Structural Composites (2003) *Fiber-bond Procedures*. [On-line] Available from: www.edgest.com [Accessed 1 September 2003].

Instron Corporation (2003) *Applications: Tensile or Tension Testing*. [On-line] Available from: www.instron.com [Accessed 1 September 2003].

Kachlakev, D. and McCurry, D. (2000) *Testing of Full-size Reinforced Concrete Beams Strengthened with FRP Composites: Experimental Results and Design Methods Verification – Report No. FHWA-OR-00-19*. Washington, DC: United States Department of Transportation Federal Highway Administration.

Khalifa, A., Alkhrdaji, T., Nanni, A., and Lansburg, S. (1999) Anchorage of surface-mounted FRP reinforcement. *Concrete International: Design and Construction*, 21(10), pp. 49–54.

Malvar, L., Warren, G., and Inaba, C. (1995) Rehabilitation of navy pier beams with composite sheets. In: *Second FRP International Symposium, Nonmetallic (FRP) Reinforcements for Concrete Structures*, Ghent, Belgium, August, pp. 533–540.

Meier, U. and Kaiser, H. (1991) Strengthening of structures with CFRP laminates. In: *Advanced Composite Materials in Civil Engineering Structures – ASCE Specialty Conference*, Las Vegas, Nevada, February. Reston, VA: ASCE, pp. 224–232.

Montgomery, D.C. and Runger, G.C. (1994) *Applied Statistics and Probability*. 1st edn. New York: John Wiley & Sons.

Nakamura, M., Sakai, H., Yagi, K., and Tanaka, T. (1996) Experimental studies on the flexural reinforcing effect of carbon fiber sheet bonded to reinforced concrete beam. In: *Proceedings of the 1st International Conference on Composites in Infrastructure, ICCI '96*, Department of Civil Engineering and Engineering Mechanics, University of Arizona, Tucson, January. Netherlands: Springer, pp. 760–773.

Norris, T., Saadatmanesh, H., and Ehsani, M. (1997) Shear and flexural strengthening of RC beams with carbon fiber sheets. *Journal of Structural Engineering*, 123(7), pp. 903–911.

Premier Corrosion Protection – Tampa (2008) *Concrete Repair* [Online images] Available at: <http://www.premier-florida.com/Concrete/Concrete.html> [Accessed 15 August 2008].

Restrepo, J. and DeVino, B. (1996) Enhancement of the axial load-carrying capacity of reinforced concrete columns by means of fiberglass epoxy jackets. In: *Proceedings of the Advanced Composite Materials in Bridges and Structures II*, Montreal, Quebec, August, pp. 547–553.

Ritchie, P.A., Thomas, D.A., Lu, L.-W., and Connelly, G.M. (1991) External reinforcement of concrete beams using fiber reinforced plastics. *ACI Structural Journal*, 88(4), pp. 490–499.

Ross, C.A., Jerome, D.M., Tedesco, J.W., and Hughes, M.L. (1999) Strengthening of reinforced concrete beams with externally bonded composite laminates. *ACI Structural Journal*, 96(2), pp. 212–220.

Saadatmanesh, H. and Ehsani, M.R. (1991) RC beams strengthened with GFRP plates, I: Experimental study. *Journal of Structural Engineering – ASCE*, 117(11), pp. 3417–3433.

Sharif, A., Al-Sulaimani, G.J., Basunbul, I.A., Baluch, M.H., and Ghaleb, B.N. (1994) Strengthening of initially loaded reinforced concrete beams using FRP plates. *ACI Structural Journal*, 91(2), pp. 160–168.

Structural Polymer Systems Limited (2001) *Guide to Composites* [CD-ROM]. Isle of Wight, UK.

# 5 Flexure

## Reinforced concrete

### 5.1 Introduction

In reinforced concrete beams, the compressive and tensile forces are resisted by concrete and reinforcement, respectively. In almost all cases, repair and rehabilitation are carried out to increase the tensile force capacity or to result in a larger moment capacity of the existing beam. Bonding steel plates to the tension face with epoxy was a popular repair method for more than two decades. However, the use of high-strength composites has become popular because of their low weight and their resistance to corrosion (ACI Committee 440, 1996).

This chapter deals with the procedures to analyze the reinforced concrete and strengthened beams and to estimate the required extra reinforcement. Even though excellent books are available for reinforced concrete design, flexural behavior of beams is explained briefly to provide a continuity in the thought process, especially in the area of tensile force transfer from extra reinforcement to the beam. The chapter is divided into sections dealing with the following major topics:

- typical load–deflection behavior of reinforced concrete beams;
- section analysis for working loads;
- section analysis and moment capacity when steel yields;
- section analysis at failure (ultimate load);
- repair methods as related to load transfer mechanism;
- section analysis of the strengthened beam at working load;
- section analysis of the strengthened beam at failure;
- serviceability considerations;
- recent developments in materials and methods of rehabilitation.

### 5.2 Load–deflection behavior of a typical reinforced concrete beam

The load–deflection behavior of reinforced concrete beams can be divided into three parts consisting of (a) precracked, (b) working load, and (c) post-yielding regions, as shown in Figure 5.1. At the initial stages of loading,

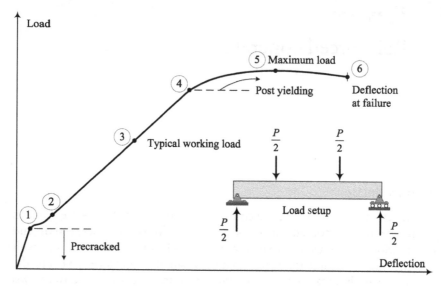

*Figure 5.1* Load–deflection behavior of a typical reinforced concrete beam.

concrete resists both compression and tension forces. When the tensile strain in the extreme fiber reaches between 0.0002 and 0.0003, the concrete starts to crack and the flexural stiffness decreases rapidly. If the instrumentation is sensitive, the rapid increase in deflection at the onset of cracking can be observed during the experimental testing. This transition occurs between points 1 and 2, as shown in Figure 5.1. Once the tension zone concrete cracks, its tensile force resistance becomes negligible. The tension force due to external load is primarily carried by reinforcement. The coupling of tension force carried by reinforcement and the compression force carried by concrete is achieved by shear through uncracked concrete.

The region between points 2 and 4 in Figure 5.1 is considered the post-cracking region. This region terminates at point 4 when reinforcement, typically mild steel, starts to yield. In almost all cases, the working load lies in this region. In other words, part of the reinforced concrete beam is cracked under load. The beam sections near simple supports or inflection points in continuous beams could still be uncracked because of lower moments.

Once the steel starts to yield, the deflection increases rapidly with very little increase in load (moment). The beam could fail by crushing of concrete or fracture of steel. In most cases, failure occurs by crushing of concrete, because the strain capacity of steel is very high. In some cases, the beam may not fail at maximum load. The process called strain softening can occur if concrete could sustain large strains because of confinement.

The increase in deflection between yielding and failure – between points 4 and 6 in Figure 5.1 – defines the ductility of the beam. If the beam is over-reinforced or concrete fails before yielding of steel, the ductility and the impending warning of failure becomes negligible. Therefore, most codes of practice around the world restrict the amount of reinforcement to ascertain yielding of steel before failure. This is achieved by limiting the reinforcement ratio to a fraction of the balanced reinforcement ratio. At the balanced reinforcement ratio, crushing of concrete and yielding of steel occur simultaneously. The readers can refer to textbooks on reinforced concrete (Limbrunner and Aghayere, 2007; MacGregor and Wight, 2005; Nawy, 2009; Setareh and Darvas, 2007; Wang *et al.*, 2007) for a more detailed discussion.

The section analysis for the various stages of loading is presented in the following sections. The discussion is very concise and the reader should refer to reinforced concrete textbooks for a more thorough understanding.

## 5.3 Section analysis: precracked reinforced concrete

A typical beam cross-section is shown in Figure 5.2. The section has both tension and compression reinforcement. If there is no compression reinforcement, substitute $A'_s = 0$. As mentioned in Section 5.2, both concrete and steel contribute in tension. If the assumptions of classical bending theory are assumed valid, the section can be analyzed using strength of materials principles.

For a given strain, the steel generates more stress than the concrete. Therefore, the area of steel can be transformed to equivalent concrete area using a modular ratio, $E_s/E_c$, where $E_s$ is the Young's modulus of steel and $E_c$ is the Young's modulus of concrete. However, for simplicity, the contribution from steel at the precracking region can be neglected. Based on this assumption, the gross section analysis leads to the following equations for the computation of various parameters.

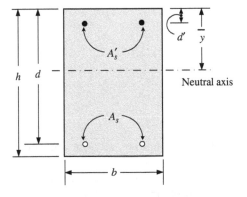

*Figure 5.2* Typical reinforced concrete beam cross-section (uncracked).

The depth of neutral axis, $kd$, is computed using the principle that the first moment of area about the neutral axis is zero. For rectangular section:

$$\bar{y} = \frac{h}{2} \tag{5.1}$$

Moment of inertia:

$$I_g = \frac{bh^3}{12} \tag{5.2}$$

Stress at extreme compression or tension fiber:

$$f_c \text{ or } f_t = \frac{M}{I_g} \times \bar{y} \text{ or } \frac{M}{I_g} \times (h - \bar{y}) \tag{5.3}$$

where $M$ is the moment applied at the cross-section.
Strain:

$$\varepsilon_{c \text{ or } t} = \frac{f_c}{E_c} \text{ or } \frac{f_t}{E_c} \tag{5.4}$$

Stress in tension reinforcement:

$$f_s = \frac{M}{I_g} \times (d - \bar{y}) \times \frac{E_s}{E_c} \tag{5.5}$$

Note that for the same strain, steel generates a larger stress:
Strain in tension reinforcement:

$$\varepsilon_s = \frac{f_s}{E_s} \tag{5.6}$$

Stress and strain in compression steel:

$$f_s' = \frac{M}{I_g}(\bar{y} - d') \times \frac{E_s}{E_c} \tag{5.7}$$

$$\varepsilon_s' = \frac{f_s'}{E_s} \tag{5.8}$$

Typical cross-sections other than the rectangular section are shown in Figure 5.3.

The most popular sections are T and double-T. L-beams are used as ledger beams to support slabs or other beams. Box girders and I-sections are typically used for prestressed beams. The equations presented below focus only on

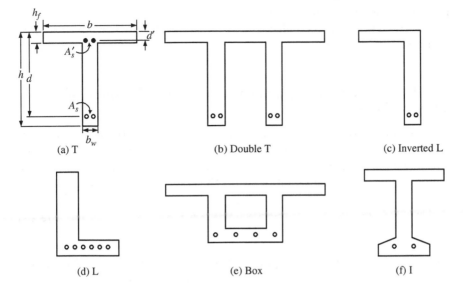

*Figure 5.3* Typical beam cross-sections.

T beams. Equations for I-sections are presented in Chapter 6 on prestressed concrete beams. Other cross-sections can be converted to equivalent T or I-sections.

In integrated construction, the width of flange, $b$, is typically restricted to (1) actual width, (2) $16h_f + b_w$, or (3) a fraction of the beam span. Appropriate code should be referred for the analysis and design. Assuming $b$ is known, the following are the equations for stress and strain computations.

Compute the depth of neutral axis using:

$$\frac{b(\overline{y})^2}{2} - (b - b_w)\frac{(\overline{y} - h_f)^2}{2} = \frac{b_w(h - \overline{y})^2}{2} \tag{5.9}$$

If $kd$ is less than $h_f$, then:

$$\frac{b(\overline{y})^2}{2} = \frac{b_w(h - \overline{y})^2}{2} + (b - b_w) \times \frac{(h_f - \overline{y})^2}{2} \tag{5.10}$$

Moment of inertia:

$$I_g = \frac{b(\overline{y})^3}{3} - (b - b_w)\frac{(\overline{y} - h_f)^3}{3} + \frac{b_w(h - \overline{y})^3}{3} \tag{5.11}$$

If $kd$ is less than $h_f$, then:

$$I_g = \frac{b(\overline{y})^3}{3} + \frac{b_w(h - \overline{y})^3}{3} + (b - b_w)\frac{(h_f - \overline{y})^3}{3} \tag{5.12}$$

Equations for stresses and strains are the same as Equations (5.3)–(5.8).

The analysis of rectangular and T-sections are further illustrated in Examples 5.1 and 5.2.

## 5.4   Estimation of cracking moment and flexural stiffness of uncracked section

The equations presented in Section 5.3 can be used to estimate the moment at first crack or cracking moment, $M_{cr}$, and flexural stiffness, $E_c I_g$. These two parameters are useful for further analysis involving deflections and strain computations. Based on Equation (5.3), the cracking moment capacity is:

$$M_{cr} = \frac{I_g}{(h - \overline{y})} f_r \tag{5.13}$$

where $f_r$ is the modulus of rupture. This value can be estimated using the compressive strength, $f_c'$. For normal-weight concrete:

$$f_r = 7.5\sqrt{f_c'} \tag{5.14}$$

Before cracking, the deflection of beams can be estimated using flexural stiffness of uncracked section, $E_c I_g$. The value of $E_c$ can be estimated using:

$$E_c = 57{,}000\sqrt{f_c'} \tag{5.15}$$

for normal-weight concrete. Both Equations (5.14) and (5.15) are based on the recommendations of the ACI code (2005). The deflection, $\delta$, for a given loading and support condition, can be expressed as:

$$\delta = \frac{PL^n}{\alpha E_c I_g} \tag{5.16}$$

Constants $n$ and $\alpha$ will depend on the support and loading conditions. For example, for a simply supported beam of span $L$ with central load $P$:

$$\delta = \frac{PL^3}{48 E_c I_g} \tag{5.17}$$

and for a simply supported beam with uniformly distributed load, $q$, per unit length:

$$\delta = \frac{5qL^4}{384 E_c I_g} \tag{5.18}$$

## 5.5   Computation of cracking moment and deflection at cracking load

*Example 5.1: Cracking moment of rectangular beam*
Estimate the cracking moment for the rectangular beam shown in Figure 5.4. In addition, estimate the deflection at the onset of cracking. The compressive strength of concrete, $f'_c$, is 5500 psi. The tension reinforcement consists of three ASTM No. 3 bars.

Solution:

Compressive strength, $f'_c = 5500 \, \text{psi}$

Modulus of rupture, $f_r = 7.5\sqrt{f'_c} = 556 \, \text{psi}$

Young's modulus of concrete, $E_c = 57{,}000\sqrt{f'_c} = 4.2 \times 10^6 \, \text{psi}$

Depth of neutral axis, $\overline{y} = \dfrac{h}{2} = \dfrac{12}{2} = 6 \, \text{in.}$

Gross moment of inertia, $I_g = \dfrac{bh^3}{12} = \dfrac{10 \times 12^3}{12} = 1440 \, \text{in.}^4$

Cracking moment, $M_{cr} = \dfrac{I_g}{(h - \overline{y})} f_r$

$\qquad = \dfrac{1440}{(12 - 6)} \times 556$

$\qquad = 133{,}440 \, \text{in.-lb}$

Beams with similar geometry tested at the University of Sherbrooke (M'Bazaa *et al.*, 1996) had an average cracking moment of 111,880 in.-lb. If their exact dimensions of 7.9 and 11.8 in. are used for $b$ and $h$, respectively, the cracking moment is 125,293 in.-lb, resulting in an error of 12%.

For the load setup shown, the deflection is:

$$\delta = \frac{P(a)(3L^2 - 4a^2)}{48E_c I_g}$$

where $P$ is the total load.

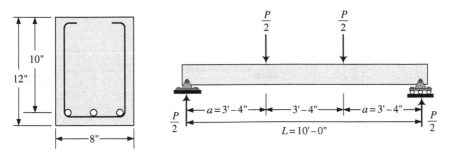

*Figure 5.4* Beam geometry and load setup for Example 5.1.

Flexural stiffness, $E_c I_g = 4.2 \times 10^6 \times 1440$

$$= 6.05 \times 10^9 \text{ lb-in.}^2$$

For a cracking moment of 133,440 in.-lb, cracking load is:

$$P_{cr} = \frac{133,440}{40} \times 2 = 6672 \text{ lb}$$

$$\delta = \frac{6672}{48 \times 6.05 \times 10^9} \times (40)(3 \times 120^2 - 4 \times 40^2)$$

$$= 0.034 \text{ in. (or } \sim 1 \text{ mm)}$$

This result agrees with the negligible deflection observed during the experimental testing.

*Example 5.2: Cracking moment of T-beam*
Compute the cracking moment and the corresponding deflection for a T-beam shown in Figure 5.5. The compressive strength of the concrete is 4500 psi. Reinforcement consists of two ASTM No. 9 bars.

Solution:

Compressive strength of concrete, $f_c' = 4500 \text{ psi}$

Modulus of rupture, $f_r = 7.5 \times \sqrt{f_c'} = 503 \text{ psi}$

Young's modulus of concrete, $E_c = 57,000 \times \sqrt{f_c'}$

$$= 3.8 \times 10^6 \text{ psi}$$

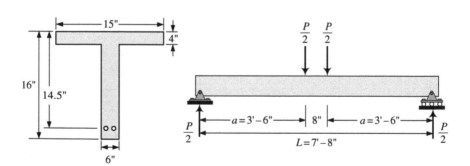

*Figure 5.5* Beam geometry and load setup for Example 5.2.

Assuming that the depth of the neutral axis, $\overline{y}$, falls below the flange thickness, then:

$$\frac{b(\overline{y})^2}{2} - (b - b_w)\frac{(\overline{y} - h_f)^2}{2} = \frac{b_w(h - \overline{y})^2}{2}$$

$$\frac{15(\overline{y})^2}{2} - (15 - 6)\frac{(\overline{y} - 4)^2}{2} = \frac{6(16 - \overline{y})^2}{2}$$

$$\overline{y} = 6.4 \text{ in.}$$

$$I_g = \frac{b(\overline{y})^3}{3} - (b - b_w)\frac{(\overline{y} - h_f)^3}{3} + \frac{b_w(h - \overline{y})^3}{3}$$

Moment of inertia:
$$= \frac{15(6.4)^3}{3} - (15 - 6)\frac{(6.4 - 4)^3}{3} + \frac{6(16 - 6.4)^3}{3}$$

$$= 3039 \text{ in.}^4$$

Cracking moment capacity, $M_{cr} = \dfrac{I_g}{(h - \overline{y})}f_r = \dfrac{3039}{(16 - 6.4)} \times 503$

$$= 159,214 \text{ in.-lb}$$

The experimental result of about 168,000 in.-lb (De Lorenzis and Nanni, 2001) is about 5% higher than the estimated value.

The cracking load, $P_{cr} = \dfrac{M_{cr}}{42} \times 2$

$$= 7582 \text{ lb}$$

Deflection, $\delta = \dfrac{7582(42)(3 \times 92^2 - 4 \times 42^2)}{48 \times 3.8 \times 10^6 \times 3039}$

$$= 0.011 \text{ in.}$$

## 5.6 Cracked section: linear elastic analysis

This section deals with the analyses for stresses, strains, and flexural stiffness of beams for moments greater than the cracking moment, $M_{cr}$, but less than the yield moment, $M_y$. The load–deflection curve for this range of loading is typically linear as shown in Figure 5.1, segments 2 to 4.

Since both concrete and steel are linearly elastic, and assumptions of classical bending theory are valid, equations similar to the ones presented in Section 5.3 are valid. The major differences are (i) the contribution of steel is taken into account and (ii) the tensile contribution of concrete is neglected. Typical strain and stress distributions for a rectangular section are shown in Figure 5.6.

The depth of the neutral axis, $kd$, can be computed using transformed section analysis. The steel area, $A_s$, is transformed to equivalent concrete area

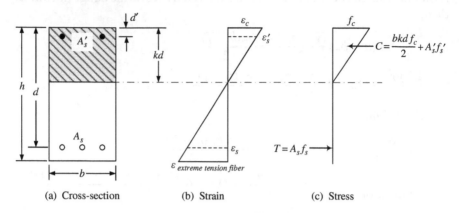

(a) Cross-section          (b) Strain          (c) Stress

*Figure 5.6* Typical stress and strain distributions for elastic analysis of rectangular section.

by multiplying $A_s$ by the ratio $E_s/E_c$, which is called the modular ratio, $n$. Once the steel area is transformed, using the first moment of area:

$$\frac{b\,(kd)^2}{2} + (n-1)\,A'_s\,(kd-d') = nA_s\,(d-kd) \tag{5.19}$$

Note that the left side provides the first moment of the compression side in which the equivalent compression steel area is taken as $(n-1)\,A'_s$. Since steel replaces concrete, $n$ is changed to $(n-1)$. On the tension side, concrete does not provide any contribution and, therefore, only the steel contribution, represented by $nA_s$, is present. If there is no compression reinforcement, $A'_s$ can be simply assumed to be zero.

Once the depth of neutral axis is known, the moment of inertia of the cracked section, $I_{cr}$, can be computed using the following equation:

$$I_{cr} = \frac{b\,(kd)^3}{3} + nA_s\,(d-kd)^2 + (n-1)\,A'_s\,(kd-d')^2 \tag{5.20}$$

For any given moment, $M$, maximum concrete stress is given by:

$$f_c = \frac{M}{I_{cr}}kd \tag{5.21}$$

Stress in tension steel is:

$$f_s = \frac{M}{I_{cr}}\,(d-kd)\,n \tag{5.22}$$

and stress in compression steel is:

$$f_s' = \frac{M}{I_{cr}} (kd - d') \, n \tag{5.23}$$

The corresponding strains in concrete, as well as tension and compression steel, can be computed using the following equations:

$$\varepsilon_c = \frac{f_c}{E_c} \tag{5.24}$$

$$\varepsilon_s = \frac{f_s}{E_s} \tag{5.25}$$

$$\varepsilon_s' = \frac{f_s'}{E_s} \tag{5.26}$$

The average extreme tension fiber strain can be computed using similar triangles.

Referring to Figure 5.6, the strain in extreme tension fiber is:

$$\varepsilon = \varepsilon_c \left( \frac{h - kd}{kd} \right) \tag{5.27}$$

The strain in the extreme tension fiber is important because strengthening systems are applied to existing structures that have some loads present. Therefore, when the strain in the strengthening composite sheet (or bar) is zero, there is a certain stress in concrete and original reinforcement, while a certain strain is present at the extreme tension face. When the analysis of strengthened beams is carried out, this difference in strains should be considered. Even if some of the loads are removed during repair, resulting in external moments less than the cracking moment of the original beam, cracked section analysis should be used for computations because once the section cracks, tension contribution of the concrete becomes negligible. The only exception is if the original beam did not crack under service loads, which is a very rare occurrence.

For T-sections, the same principles apply. The following discussion provides the appropriate equations. Refer to Figure 5.3 for an explanation of the notation.

If the depth of the neutral axis is less than the flange thickness, $h_f$, then:

$$\frac{b(kd)^2}{2} + (n - 1) A_s' (kd - d') = nA_s (d - kd) \tag{5.28}$$

Note that Equation (5.28) is the same as the equation for a rectangular beam, (5.19). Since the tension force contribution of concrete is neglected, if the neutral axis falls within the flange, the beam behaves in a similar way to a rectangular beam with width $b$, and depth $d$.

If the depth of the neutral axis is greater than the flange thickness, then:

$$\frac{b(kd)^2}{2} - (b - b_w)\frac{(kd - h_f)^2}{2} + (n - 1)\, A_s'\, (kd - d') = nA_s\, (d - kd)$$

$$(5.29)$$

Equations (5.28) and (5.29) are valid for inverted-L, double-T, I, and box sections as long as the neutral axis falls above the bottom flange of the I or box sections. For double-T, the web width, $b_w$, should be taken as the total width of two webs, and the same is true for box sections. The L (ledger) beams will behave as rectangular beams as long as the neutral axis falls above the bottom flange.

If the neutral axis falls within the flange thickness, then:

$$I_{cr} = \frac{b\,(kd)^3}{3} + nA_s\, (d - kd)^2 + (n - 1)\, A_s'\, (kd - d')^2 \qquad (5.30)$$

which is the same as Equation (5.20).

If the neutral axis falls below flange thickness, then:

$$I_{cr} = \frac{b(kd)^3}{3} - (b - b_w)\frac{(kd - h_f)^3}{3} + nA_s(d - kd)^2 + (n - 1)A_s'(kd - d')^2$$

$$(5.31)$$

Equations for stresses and strains are the same as Equations (5.21)–(5.27).

Equations presented in this section are not valid if the steel yields or the stresses in the concrete exceed the linear range. For normal-strength concrete, stress–strain behavior can be assumed linear up to $0.45f_c'$. For high-strength concrete, the linear range extends to about $0.7f_c'$.

## 5.7   Flexural stiffness in the working load range

As in the case of uncracked section, flexural stiffness of cracked section is $E_c I_{cr}$. However, since parts of the beams do not crack owing to the lower moments, and the sections between cracks contribute to flexural stiffness, the average stiffness is greater than $E_c I_{cr}$. This average stiffness can be estimated using the ACI code guidelines (2005). The effective moment of inertia, $I_e$, is computed as a function of the cracking moment, $M_{cr}$, and the maximum moment, $M_a$.

$$I_e = I_{cr} + \left(I_g - I_{cr}\right)\left(\frac{M_{cr}}{M_a}\right)^3 \qquad (5.32)$$

The magnitude of the second term decreases cubically with respect to the maximum moment. Even though Equation (5.32) was developed for uniformly distributed load, it can be used for other types of load as a good approximation. However, if only concentrated loads are present, the equation should be used with caution.

The effective stiffness, $E_c I_e$, can be used for deflection computations and average strains at the extreme tension face.

## 5.8 Computation of stresses and strains in the post-crack, preyielding stage

*Example 5.3: Stresses and strains in a rectangular beam*
Using the details of Example 5.1, compute the maximum stresses and strains in concrete and steel for a load, $P = 8000\,\text{lb}$. Also, compute the mid-span deflection. Assume a modulus of elasticity of $29 \times 10^6$ psi for steel.

Solution:

Using the details provided in Example 5.1:

Width, $b = 8\,\text{in.}$

Depth, $d = 10\,\text{in.}$

Thickness, $h = 12\,\text{in.}$

Area for an ASTM No. 3 bar is $0.11\,\text{in.}^2$:

Area of steel, $A_s = 3 \times 0.11 = 0.33\,\text{in.}^2$

Computation of the depth of the neutral axis:

$$\frac{b(kd)^2}{2} = nA_s(d - kd)$$

$E_c = 4.2 \times 10^6$ psi (Example 5.1)

$E_s = 29 \times 10^6$ psi

$$n = \frac{E_s}{E_c} = \frac{29}{4.2} = 6.9$$

Therefore:

$$\frac{8}{2}(kd)^2 = 6.9 \times 0.33\,(10 - kd)$$

$$kd = 2.12\,\text{in.}$$

Moment of inertia of the cracked section:

$$I_{cr} = \frac{1}{3}b(kd)^3 + nA_s(d - kd)^2$$

$$= \frac{1}{3} \times 8 \times 2.12^3 + 6.9 \times 0.33(10 - 2.12)^2 = 166.8 \text{ in.}^4$$

Maximum moment, $\frac{P}{2} \times 40 = 4000 \times 40 = 160{,}000 \text{ in.-lb}$

Maximum stress in concrete, $\frac{160{,}000}{166.8} \times 2.12 = 2033 \text{ psi}$

The compressive strength of concrete, $f_c' = 5500 \text{ psi}$.
The maximum stress is $0.37f_c'$.
Hence, the linear elastic behavior of concrete can be assumed correct.

Maximum steel stress, $\frac{160{,}000}{166.8}(10 - 2.12) \times 6.9 = 52{,}154 \text{ psi}$

Assuming the steel has a yield stress of 60,000 psi, the behavior of the steel is also linearly elastic.

Maximum strain in steel is $\frac{52{,}154}{29 \times 10^6} = 0.0018$.

Maximum compressive strain in concrete is $\frac{2033}{4.2 \times 10^6} = 0.00048$.

Average strain at the extreme tension face is

$$0.00048\left(\frac{12.0 - 2.12}{2.12}\right) = 0.0023$$

If high-strength composites are attached, with the beam carrying a load of 8000 lb, the strain in the composite will be zero when the strains in the concrete and steel are 0.00048 and 0.0018, respectively. Only a small amount of extra load can be applied before the steel yields. Therefore, some load has to be released when applying the composites.

Computation of mid-span deflection:

Cracking moment, $M_{cr} = 133{,}440 \text{ in.-lb}$

Maximum moment, $M_a = 160{,}000 \text{ in.-lb}$

Moment of inertia of gross section, $I_g = 1440 \text{ in.}^4$

Moment of inertia of cracked section, $I_{cr} = 166.8 \text{ in.}^4$

Effective moment of inertia, $I_e = I_{cr} + (I_g - I_{cr})\left(\frac{M_{cr}}{M_a}\right)^3$

$$= 166.8 + (1440 - 166.8)\left(\frac{133{,}440}{160{,}000}\right)^3$$

$$= 905.3 \text{ in.}^4$$

$$\text{Deflection, } \delta = \frac{8000(40)(3 \times 120^2 - 4 \times 40^2)}{48 \times 4.2 \times 10^6 \times 905.3}$$

$$= 0.073 \, \text{in.}$$

This deflection value is close to the experimental value reported by M'Bazaa *et al.* (1996).

*Example 5.4: Stresses and strains in T-beam*
Using the details of Example 5.2, compute the maximum stresses, strains, and the mid-span deflection for a total load of 50,000 lb. Assume the modulus of elasticity of steel as $29 \times 10^6$ psi.

Solution:

Using the details of Example 5.2,

Flange width, $b = 15 \, \text{in.}$

Flange thickness, $h_f = 4 \, \text{in.}$

Web width, $b_w = 6 \, \text{in.}$

Depth, $d = 14.5 \, \text{in.}$

Thickness, $h = 16 \, \text{in.}$

Area of steel, $A_s = 2 \times 1.34 = 2.68 \, \text{in.}^2$

Assuming the depth of the neutral axis falls within the flange, then:

$$\frac{b(kd)^2}{2} = nA_s \left( d - kd \right)^2$$

$$n = \frac{29 \times 10^6}{3.8 \times 10^6} = 7.6$$

$$\frac{15(kd)^2}{2} = 7.6 \times 2.68 \left( 14.5 - kd \right)$$

$kd = 5 \, \text{in.} >$ Flange thickness of 4 in.

Therefore recompute the depth of the neutral axis using:

$$\frac{b(kd)^2}{2} - \frac{(b - b_w)(kd - h_f)^2}{2} = nA_s \left( d - kd \right)$$

$$\frac{15(kd)^2}{2} - \frac{(15 - 6)(kd - 4)^2}{2} = 7.6 \times 2.68 \left( 14.5 - kd \right)$$

$kd = 5.2 \, \text{in.}$

Moment of inertia of cracked section:

$$I_{cr} = \frac{b(kd)^3}{3} - \frac{(b - b_w)(kd - b_f)^3}{3} + nA_s(d - kd)^2$$

$$= \frac{15(5.2)^3}{3} - \frac{(15 - 6)(5.2 - 4)^3}{3} + 7.6 \times 2.68(14.5 - 5.2)^2$$

$$= 2460\,\text{in.}^4$$

Note: In Example 5.3, when the section cracked, the moment of inertia reduced from 1440 to 166.8 in.⁴, whereas in this example, the change was from 3039 to 2460 in.⁴. In the first example, the reinforcement area was small and, hence, the depth of the neutral axis was very small. This resulted in a loss of a large area of cross-section below the neutral axis (9.88 out of 12 in. was lost). In the second example, since the area of steel was high, only a small part of the area was lost. Most of the compression area in the flange was preserved, resulting in a much higher retention of uncracked area and the moment of inertia.

Maximum moment, $M_a = \dfrac{P}{2} \times 42$

$$= \left(\frac{50,000}{2}\right) \times 42 = 1,050,000\,\text{in.-lb}$$

Maximum stress in concrete, $\dfrac{1,050,000}{2460} \times 5.2 = 2218\,\text{psi}$

Maximum stress in steel, $\dfrac{1,050,000}{2460} \times (14.5 - 5.2) \times 7.6 = 30,167\,\text{psi}$

Maximum compressive strain in concrete, $\dfrac{2218}{3.8 \times 10^6} = 0.00058.$

Maximum strain in steel, $\dfrac{30,167}{29 \times 10^6} = 0.00104.$

Average strain in the extreme tension face,

$$0.00058 \left(\frac{16 - 5.2}{5.2}\right) = 0.0012.$$

Computation of maximum (mid-span) deflection:

$M_{cr} = 159,214\,\text{in.-lb}$
$M_a = 1,050,000\,\text{in.-lb}$

Effective moment of inertia, $I_e = 2460 + (3039 - 2460)\left(\dfrac{159,214}{1,050,000}\right)^3$

$$= 2462\,\text{in.}^4$$

Deflection, $\delta = \dfrac{50,000(42)(3 \times 92^2 - 4 \times 42^2)}{48 \times 3.8 \times 10^6 \times 2642}$

$$= 0.086\,\text{in.}$$

The experimental deflection is about 0.08 in. (De Lorenzis and Nanni, 2001).

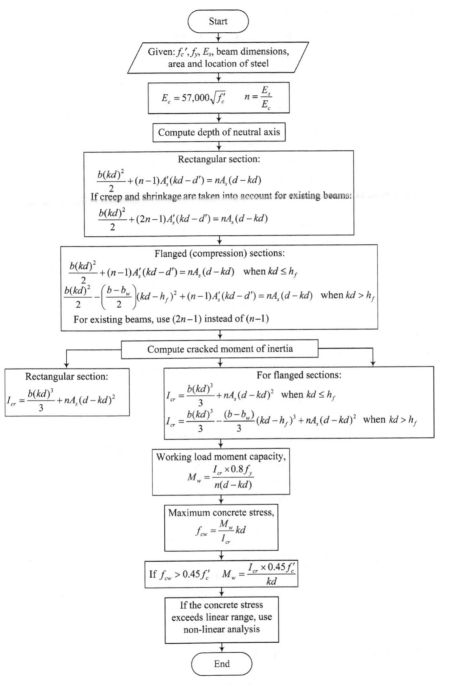

Figure 5.7 Flowchart for computation of moment capacity at working loads: unstrengthened reinforced concrete beams.

## 5.9   Computation of moment capacities at yielding of steel or at concrete stress of $0.45f_c'$

The moment capacity of the beam when the steel reinforcement starts yielding (yield moment) is very important for the following reasons:

- The moment due to the working loads should not exceed 80% of the yield moment of the strengthened section. This possibility could occur when lightly reinforced beams are strengthened with a large amount of composite reinforcement.
- The contribution of additional composite reinforcement becomes significant after yielding of steel. In typical reinforced concrete beams, once the steel yields, the rate of increase of moment becomes negligible because there is very little additional tension force until the steel reaches the strain-hardening stage. Since composite contribution remains linear up to failure, strengthened beams provide considerable strength increase beyond the yielding of steel.

The maximum limit on concrete stress of $0.45f_c'$ is based on the ACI code (2005). The behavior of concrete can be assumed linearly elastic up to this stress limit. At yielding of steel, if the complete stress exceeds $0.45f_c'$, then this stress limit will control the maximum allowable working loads. In this section, only the reinforcement is considered. The contribution of composite reinforcement is discussed in Section 5.15.

The equations developed for preyield conditions are valid up to the onset of yielding of steel, provided that the concrete remains in the linear-elastic range. If the maximum concrete stress is less than $0.45f_c'$, this assumption is valid. If the concrete stress exceeds this limit, the allowable working moment should be recomputed using this stress limit. In this case, the steel remains linearly elastic because the stress is less than yield strength.

Since computation of yield moment is an important step in the design process, the calculation sequence is shown in a flowchart (Figure 5.7). Numerical examples are presented in the next section.

## 5.10   Computation of working load moment capacity

*Example 5.5: Working load moment capacity of rectangular beam*
Using the details and results of Examples 5.1 and 5.3, compute the working load moment capacity, $M_w$, for the rectangular cross-section of Example 5.1, assuming the yield stress of steel is 60,000 psi.

Solution:

Using the results of Example 5.3:

Depth of neutral axis, $kd = 2.12$ in.

Moment of inertia of cracked section, $I_{cr} = 166.8$ in.[4]

$$M_w = \frac{166.8 \times 0.8 \times 60{,}000}{(10 - 2.12) \times 6.9} \text{ in.-lb}$$

$$= 147{,}252 \text{ in.-lb}$$

Maximum concrete stress (when steel yields):

$$f_{cy} = \frac{147{,}252}{166.8} \times 2.12 = 1871 \text{ psi}$$

$$\frac{1871}{5500} = 0.34f_c' < 0.45f_c'$$

Therefore working load moment capacity is 147,252 in.-lb.

$$\text{Yield moment,} \quad \frac{M_w}{0.8} = \frac{147{,}252}{0.8} = 184{,}065 \text{ in.-lb}$$

Experimentally observed yield moment was 184,000 in.-lb (M'Bazaa *et al.*, 1996).

### *Example 5.6: Working load moment capacity of T-beam*
Compute the working load moment capacity, $M_w$, for the T-beam of Example 5.4, assuming the yield stress of steel = 60,000 psi.

Solution:

Using the results of Example 5.4:

Depth of neutral axis, $kd = 5.2$ in.

Moment of inertia of cracked section, $I_{cr} = 2460$ in.[4]

$$\text{Working load moment capacity, } M_w = \frac{2460 \times 0.8 \times 60{,}000}{(14.5 - 5.2) \times 7.6}$$

$$= 1{,}670{,}628 \text{ in.-lb}$$

$$\text{Maximum concrete stress, } f_{cy} = \frac{1{,}670{,}628}{2460} \times 5.2 = 3531 \text{ psi}$$

$$= \frac{3532}{4500} = 0.78f_c' > 0.45f_c' M_w$$

$$= \frac{2460}{5.2} \times 0.45 \times 4500$$

$$= 957{,}980 \text{ in.-lb}$$

The experimental testing showed a considerable nonlinear behavior (De Lorenzis and Nanni, 2001). Eighty percent of the experimentally observed yield moment is about 1,176,000 in.-lb, which is slightly higher than 957,980 in.-lb, but much less than 1,670,628 in.-lb. The experimental value is lower because of the nonlinear behavior of concrete.

## 5.11   Section analysis at failure, nominal resisting moment, $M_n$

Almost all the codes around the world permit the design of reinforced concrete members based on strength design. A large number of excellent text-books and design guides are available for analysis and design (Limbrunner and Aghayere, 2007; MacGregor and Wight 2005; Nawy, 2009; Setareh and Darvas, 2007; Wang *et al.*, 2007). This section provides a short summary of the analysis of rectangular and flanged sections. It is assumed that the reader is familiar with the subject.

### 5.11.1   Basic assumptions and guidelines

The following assumptions and guidelines are chosen based on the recommendation of ACI code (2005) (Refer to Figure 5.6).

- The strain distribution across the thickness is linear.
- Maximum strain in concrete at failure is 0.003.
- The stress distribution of concrete can be assumed rectangular with an average stress of $0.85f_c'$ and a depth of $a$.
- The depth of stress block, $a = \beta_1 \times$ depth of neutral axis, $c$.
- $\beta_1 = 0.85$ if $f_c' \le 4000$ psi

  - $= 0.65$ if $f_c' \ge 8000$ psi
  - $= 0.85 - 0.05 \left( \frac{f_c' - 4000}{1000} \right)$ if $4000 < f_c' < 8000$ psi.

- The reinforcement ratio, $\rho = A_s/bd$, should satisfy both minimum and maximum requirements. For flanged sections, use $b_w$.
  This inequality is based on $E_s = 29 \times 10^6$ psi.
  For sections with compression reinforcement:

$$\rho \le 0.75\overline{\rho_b} + \rho' \frac{f_s'}{f_y}$$

  where $\rho' = A_s'/bd$ and $f_s'$ is the stress in compression steel at failure.
- One-way slabs can be designed as rectangular beams with a width of 12 in.
- For T-beams, the flange width is the minimum of (i) actual flange width (precast section), (ii) $(16h_f + b_w)$, (iii) span/4, or (iv) centerline spacing of webs.
- For L-beams, the flange width is the minimum of (i) actual flange width, (ii) $(6h_f + b_w)$, (iii) span/12, or (iv) half clear distance to next web + web width.
- The maximum stress in steel is $f_y$.

The depth of the neutral axis is computed using the force equilibrium equations.

For rectangular beams with tension reinforcement only:

$$0.85 f_c' ba = A_s f_y \tag{5.33}$$

$$a = \beta_1 c \tag{5.34}$$

For rectangular beams with compression reinforcement:

$$0.85 f_c' ba + A_s' f_s' = A_s f_y \tag{5.35}$$

The stress in compression steel, $f_s'$, has to be computed by trial and adjustment. First assume $f_s = f_y$ and compute $a$ and $c$, and the strain in compression steel as:

$$\varepsilon_s' = 0.003 \left( \frac{c - d'}{c} \right) \tag{5.36}$$

If $\varepsilon_s' \geq \varepsilon_y$ then $f_s' = f_y$. If $\varepsilon_s'$ is less than $\varepsilon_y$, then:

$$f_s' = E_s \varepsilon_s' \tag{5.37}$$

Use this value to recompute $a$ using Equation (5.35). Repeat the process until the values of $f_s'$ converge. For flanged sections with tension reinforcement only, assume that $c$ is less than the flange thickness, $h_f$, and use Equation (5.33). If $c$ is greater than $h_f$, then:

$$0.85 f_c' bh_f + 0.85 f_c' b_w (a - h_f) = A_s f_y \tag{5.38}$$

If compression reinforcement is present, use Equation (5.35) and the procedure similar to the rectangular beam, or if $c > h_f$, use:

$$0.85 f_c' bh_f + 0.85 f_c' b_w (a - h_f) + A_s' f_s' = A_s f_y \tag{5.39}$$

Determine $f_s'$ using the iteration procedure. Nominal moment capacity, $M_n$, is computed using moment equilibrium equations.

For rectangular section with no compression reinforcement, use:

$$M_n = 0.85 f_c' ba \left( d - \frac{a}{2} \right) \tag{5.40}$$

For rectangular beams with compression reinforcement, use:

$$M_n = 0.85 f_c' ba \left( d - \frac{a}{2} \right) + A_s' f_s' (d - d') \tag{5.41}$$

For flanged sections with no compression reinforcement, use:

$$M_n = 0.85f_c'ba\left(d - \frac{a}{2}\right) \text{ if } h_f \geq c \tag{5.42}$$

$$= 0.85f_c'\left(b - b_w\right)h_f\left(d - \frac{h_f}{2}\right) + 0.85f_c'b_wa\left(d - \frac{a}{2}\right)$$

$$\text{if } h_f < c \tag{5.43}$$

For flanged sections with compression reinforcement, use:

$$M_n = 0.85f_c'ba\left(d - \frac{a}{2}\right) + A_sf_s'\left(d - d'\right) \text{ if } c \leq h_f \tag{5.44}$$

$$= 0.85f_c'\left(b - b_w\right)h_f\left(d - \frac{h_f}{2}\right) + 0.85f_c'b_wa\left(d - \frac{a}{2}\right) + A_s'f_s'\left(d - d'\right)$$

$$\text{if } c > h_f \tag{5.45}$$

The external moment computed using factored loads, $M_u$, should be less than or equal to $\phi M_n$. For flexure, $\phi$ is taken as 0.9.
The factored load is:

$$w_u = 1.2w_D + 1.6w_L \tag{5.46}$$

where $w_D$ is the dead load and $w_L$ is the live load. Load factors for other loads such as wind loads and combinations can be found in the ACI code (2005).

The sequence of calculations for the computation of $M_n$ is presented in the form of flowcharts in Figures 5.8–5.10 for rectangular and flanged beams. The equations are repeated in the flowcharts so that the reader can follow the whole sequence without referring back to the text. As mentioned earlier, the reader is encouraged to consult the textbooks on reinforced concrete for understanding the fundamental concepts, reasons, and validations for the various assumptions.

## 5.12   Computation of nominal moment, $M_n$

*Example 5.7: Nominal moment capacity of rectangular beam*
Compute the nominal moment capacity of the rectangular beam of Example 5.1. Assume yield strength of 60,000 psi and a modulus of elasticity of $29 \times 10^6$ psi for steel.

Solution:

The computation is carried out following the sequence of steps outlined in the flowchart of Figure 5.8.

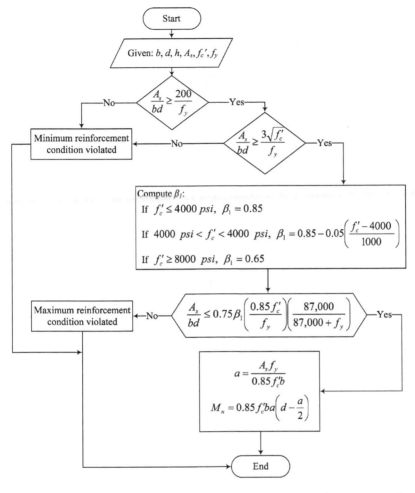

*Figure 5.8* Flowchart for computation of $M_n$: rectangular section with tension reinforcement.

Given:

$b = 8$ in.

$d = 10$ in.

$h = 12$ in.

$A_s = 0.33$ in.$^2$

$f'_c = 5500$ psi

$f_y = 60,000$ psi

$E_s = 29 \times 10^6$ psi

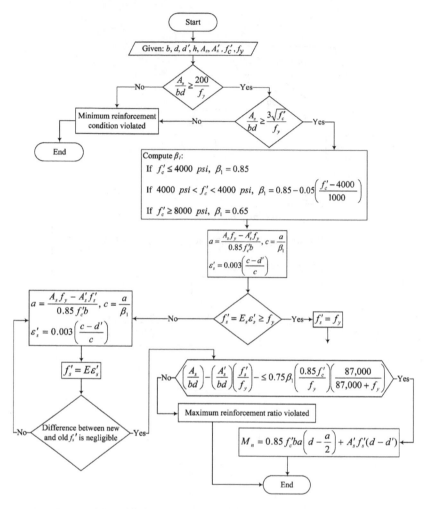

*Figure 5.9* Flowchart for computation of $M_n$: rectangular section with compression and tension reinforcement.

Check for minimum reinforcement:

$$\frac{A_s}{bd} = \frac{0.33}{8 \times 10} = 0.0041$$

$$\frac{200}{f_y} = \frac{200}{60,000} = 0.0033$$

$$\frac{3\sqrt{f_c'}}{f_y} = \frac{3\sqrt{5500}}{60,000} = 0.0037$$

$$\frac{A_s}{bd} > \frac{3\sqrt{f_c'}}{f_y} > \frac{200}{f_y}$$

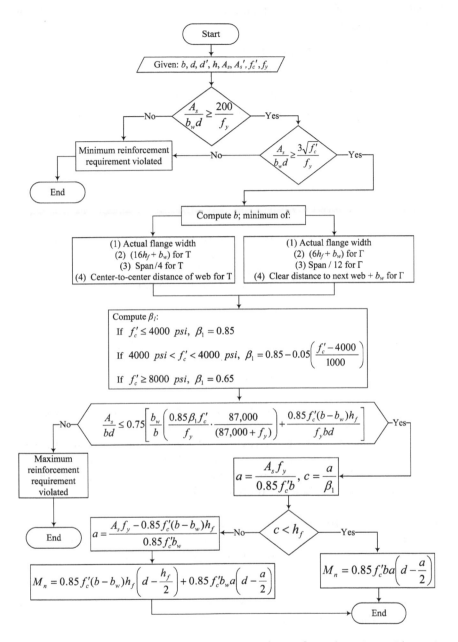

*Figure 5.10* Flowchart for computation of $M_n$: flanged section with tension reinforcement.

Therefore, the reinforcement satisfies the minimum requirement. Check for maximum reinforcement:

$$\beta_1 = 0.85 - 0.05\left(\frac{5500-4000}{1000}\right) = 0.775$$

$$0.75\beta_1\left(\frac{0.85f_c'}{f_y}\right)\left(\frac{87,000}{87,000+f_y}\right) = 0.75 \times 0.775 \left(\frac{0.85 \times 5500}{60,000}\right)$$

$$\left(\frac{87,000}{87,000+60,000}\right) = 0.0268$$

$$\frac{A_s}{bd} = 0.0041 < 0.0268$$

Therefore, the reinforcement satisfies the restriction on maximum area.

Depth of stress block, $a = \dfrac{A_s f_y}{0.85 f_c' b} = \dfrac{0.33 \times 60,000}{0.85 \times 5500 \times 8} = 0.53\,\text{in.}$

Nominal moment, $M_n = 0.85 f_c' ba\left(d - \dfrac{a}{2}\right)$

$$= 0.85 \times 5.500 \times 8 \times 0.53\left(10 - \frac{0.53}{2}\right)$$

$$= 193,967\,\text{in.-lb}$$

The experimental moment at a concrete strain of 0.003, reported by M'Bazaa *et al.* (1996), was about 230,000 in.-lb. The concrete failed at a strain of 0.015. This resulted in strain hardening of steel and a failure moment of 283,200 in.-lb. This type of behavior occurs in beams with very low reinforcement ratio. Note that the reinforcement ratio for the example is 0.0041, whereas the ratio for balanced failure is 0.0358.

### Example 5.8: Nominal moment capacity of T-beam
Compute the nominal moment capacity of the T-beam of Example 5.2. Assume a yield strength of 60,000 psi and modulus of $29 \times 10^6$ psi.

Solution:

The sequence of calculations presented in the flowchart of Figure 5.10 is followed for the computation.
   Given:

$b = 15\,\text{in.}$

$b_w = 6\,\text{in.}$

$d = 14.5\,\text{in.}$

$h_f = 4\,\text{in.}$

$A_s = 2.0\,\text{in}^2$

$f'_c = 4500\,\text{psi}$

$f_y = 60{,}000\,\text{psi}$

$E_s = 29 \times 10^6\,\text{psi}$

Check for minimum reinforcement:

$$\frac{A_s}{b_w d} = \frac{2.0}{6 \times 14.5} = 0.023$$

$$\frac{200}{f_y} = \frac{200}{60{,}000} = 0.0033$$

$$\frac{3\sqrt{f'_c}}{f_y} = \frac{3\sqrt{4500}}{60{,}000} = 0.0034$$

$$\frac{A_s}{b_w d} > \frac{3\sqrt{f'_c}}{f_y} > \frac{200}{f_y}$$

Minimum reinforcement requirement is satisfied. Check for maximum reinforcement:

$$\beta_1 = 0.85 - 0.05 \left(\frac{4500 - 4000}{1000}\right) = 0.825$$

$$= 0.75 \left[\frac{b_w}{b}\left(\frac{0.85\beta_1 f'_c}{f_y}\right)\left(\frac{87{,}000}{87{,}000 + f_y}\right) + \frac{0.85 f'_c(b - b_w)h_f}{f_y b d}\right]$$

$$= 0.75 \left[\frac{6}{15}\left(\frac{0.85 \times 8.25 \times 4500}{60{,}000}\right)\frac{87{,}000}{(87{,}000 + 60{,}000)}\right.$$

$$\left. + \frac{0.85 \times 4500(15 - 6)4}{60{,}000 \times 15 \times 14.5}\right]$$

$$= 0.75\,[0.0311 + 0.0106] = 0.0313$$

$$\frac{A_s}{b_w d} = 0.023 < 0.0313$$

Therefore, maximum reinforcement requirement is satisfied. Assuming that the depth of the neutral axis is less than the flange thickness:

Depth of stress block, $a = \dfrac{A_s f_y}{0.85 f_c' b}$

$a = \dfrac{2.00 \times 60,000}{0.85 \times 4500 \times 15} = 2.09\,\text{in.}$

Depth of the neutral axis $c = \dfrac{a}{\beta_1} = \dfrac{2.09}{0.825} = 2.46\,\text{in.} < h_f$ or $4\,\text{in.}$

Nominal moment, $M_n = 0.85 f_c' \, ba \left( d - \dfrac{a}{2} \right)$

$= 0.85 \times 4500 \times 15 \times 2.09 \times \left( 14.5 - \dfrac{2.09}{2} \right)$

$= 1,612,840\,\text{in.-lb}$

The experimental moment reported by De Lorenzis and Nanni (2001) was 1,705,200 in.-lb, resulting in an error of 5%. Note the experimental yield strength of steel was 62,000 psi instead of 60,000 psi used in the example. In addition, the compressive strength may not be exact.

*Example 5.9: Rectangular slab with tension and compression reinforcement*
Using the details provided in Figure 5.11, compute the nominal moment if compressive strength, $f_c' = 5100$ psi, yield strength of steel, $f_y = 82,000$ psi, and modulus of elasticity of steel $= 29 \times 10^6$ psi (slabs tested by Crasto *et al.*, 1996).

Solution:

Refer to flowchart of Figure 5.9 for the sequence of calculations.

Given:

$b = 18\,\text{in.}$

$d = 4.5\,\text{in.}$

$d' = 1.5\,\text{in.}$

$A_s' = 0.22\,\text{in.}^2$

$A_s = 1.24\,\text{in.}^2$

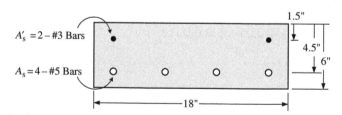

*Figure 5.11* Geometric details of beam of Example 5.9.

Check for minimum reinforcement:

$$\frac{A_s}{bd} = \frac{1.24}{18 \times 6} = 0.0115$$

$$\frac{200}{f_y} = \frac{200}{82,000} = 0.0024$$

$$\frac{3\sqrt{f_c'}}{f_y} = \frac{3\sqrt{5100}}{82,000} = 0.0026$$

$$\frac{A_s}{bd} > \frac{3\sqrt{f_c'}}{f_y} > \frac{200}{f_y}$$

The tension steel area satisfies the minimum reinforcement requirement. However, the ratio is very low. Computation of depth of neutral axis:

$$\beta_1 = 0.85 - 0.05 \left( \frac{5100 - 4000}{1000} \right) = 0.795$$

Assuming compression steel yields, depth of stress block is:

$$a = \frac{A_s f_y - A_s' f_y}{0.85 f_c' b}$$

$$= \frac{1.24 \times 82,000 - 0.22 \times 82,000}{0.85 \times 5100 \times 18} = 1.07 \, \text{in.}$$

$$c = \frac{a}{\beta_1} = 1.04 \, \text{in.}$$

The compressive reinforcement is close to the neutral axis and, hence, the stresses are negligible. The slab can be treated as a singly reinforced slab.

$$f_s' \cong 0$$

$$0.75 \beta_1 \left( \frac{0.85 f_c'}{f_y} \right) \left( \frac{87,000}{87,000 + f_y} \right)$$

$$= 0.75 \times 0.795 \left( \frac{0.85 \times 5100}{82,000} \right) \left( \frac{87,000}{87,000 + 82,000} \right)$$

$$= 0.016$$

The reinforcement ratio, $\dfrac{A_s}{bd} - \dfrac{A_s'}{bd} \left( \dfrac{f_s'}{f_y} \right) \cong 0.0115 < 0.016$

Therefore, the maximum reinforcement requirement is satisfied.

Assuming $f_s' \cong 0$, depth of stress block:

$$a = \frac{A_s f_y}{0.85 f_c' b} = \frac{1.24 \times 82,000}{0.85 \times 5100 \times 18} = 1.31 \text{ in.}$$

Nominal moment:

$$M_n = 0.85 f_c' ba \left( d - \frac{a}{2} \right) + A_s' f_s' \left( d - d' \right)$$

$$= 0.85 \times 5100 \times 81 \times 1.31 \left( 4.5 - \frac{1.31}{2} \right) + 0 = 393,386 \text{ in.-lb}$$

The experimental maximum moment was 380,000 in.-lb, so the error of prediction is 3.5%.

## 5.13   Load–deflection behavior of typical strengthened reinforced concrete beams

Typical load–deflection curves for strengthened reinforced beams are shown in Figure 5.12. Figure 5.12(a) shows the load–deflection response of a beam strengthened when no external loads are present. This curve represents the behavior of beams made in the laboratory for evaluation. Theoretically, it is feasible to reduce the stresses to zero in the field by using props. However, in most, if not all cases, it is neither practical nor economical to bring the beam to zero stress level. However, the curve in Figure 5.12(a) can be used for basic understanding.

The beam strengthened at zero load essentially behaves likes the reinforced concrete beam with extra reinforcement (Section 5.2) with the following notable differences:

- The cracking load (moment) is slightly larger due to the extra force provided by the composite.
- In the post-crack, preyielding region, between points 2 and 3 in Figure 5.12(a), the slope will be higher. The higher stiffness (less deflection) is a function of composite plate cross-sectional area and modulus of elasticity of the fiber. For example, carbon plates will provide a larger increase than glass plates. However, in the normal strengthening range (strength increase less than 50%) the stiffness increase in not significant.
- The load (moment) at which yielding of steel occurs will also increase. The increase is a again function of composite thickness and its modulus of elasticity. Computations of stiffness and the yield load are presented in the following sections.
- The contribution of composite becomes very significant in the post yielding stage, between points 3 and 4 of Figure 5.12(a). Since additional contribution of steel is zero in the yield plateau, post-yield part of the

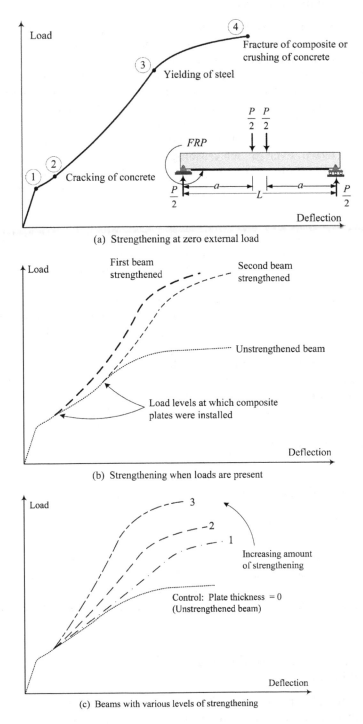

(a) Strengthening at zero external load

(b) Strengthening when loads are present

(c) Beams with various levels of strengthening

*Figure 5.12* Typical load–deflection curves of strengthened reinforced concrete beams.

curve is flat for a reinforced concrete beam. The composite strengthened beam continues to provide strength increase because the composite force contribution continues at the same level.

- Since steel yields prior to failure, failure occurs by crushing of concrete or failure of composite. Failure of composite could occur due to fracture or delamination. Both fracture and delamination are brittle but fracture failure is typically less brittle. In addition, fracture can be predicted with more certainty. Therefore, between the two, failure by fracture of composite is preferred. Even though crushing of concrete is also sudden, this failure is preferred instead of composite failure. Due to the presence of some form of confinement, crushing of concrete tends to be less brittle than composite failure.

The load–deflection response shown in Figure 5.12(b) is more typical in field applications. The composite is installed when certain external loads are present. In most cases, the beam will be in the post-crack, preyield stage. Even if some of the loads such as live loads are removed, the beams are still in cracked condition. The behavior is similar to the previous case, except that the composite has more strain capacity and, hence, can achieve larger deflections. Since composites can generate large strains compared to yield strain of steel, the loads should be reduced as much as possible during the installation.

The influence of composite thickness is shown in Figure 5.12(c). Larger plate thicknesses provide larger stiffness increase, higher load at steel yielding, higher load postyield increase, and higher failure loads. However, as the thickness increases, failure by delamination of the plate could become a possibility. This aspect is discussed further in Section 5.19.

For design (analysis) purposes, the following three stages are of importance:

(i) The level of loading at which the plates are installed
(ii) Capacity at yielding of steel
(iii) Failure or ultimate moment.

The first one influences the next two parameters. The working load should always be less than yield load. It is recommended that the increased level of working load controls the strength and, hence, is a critical parameter in design. For most cases, failure load will control the design of strengthening system.

## 5.14  Basic assumptions for the analysis of strengthened beams

Typically, additional reinforcement is bonded to the existing structures using epoxy. A number of investigations have been carried out in which aramid, carbon, and glass sheets (fabrics) have been attached to the tension faces to increase the flexural strength (ACI Committee 440, 2002). Nanni and his research team have developed a novel method of attaching composite bars to an existing structures (Nanni and Faza, 2002). For the analysis of strengthened beams, the following additional assumptions, in addition to the one made for reinforced concrete, are needed.

- There is a perfect bond between composite (plate or bar) and the beam, up to failure.
- Linear strain distribution is valid even under large curvatures.
- The behavior of the composite plate is linearly elastic up to failure.
- The stresses and strains in the original beam, at the time of installation of the composite, can be computed using cracked section elastic analysis.
- For working load analysis, the composite can be considered as an additional reinforcement.
- The composite plates are thin and the center of gravity can be assumed to be located at a distance, $h$, from the extreme compression fiber.

## 5.15  Analysis of cracked section: strengthened reinforced concrete

As mentioned in Section 5.12, in most cases, strengthening is done when the beams are at the post-cracked stage. Even if the beam is uncracked, the analysis is the same as the one presented in Section 5.3 and, hence, analysis of uncracked, strengthened beams is not presented.

Analysis of strengthened beams in the linear elastic range is similar to the analysis of unstrengthened beams. The only difference is the addition of the force contribution as shown in Figure 5.13. If the composite is installed

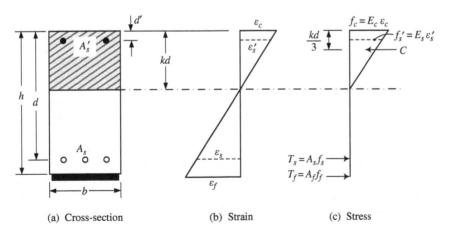

(a) Cross-section        (b) Strain        (c) Stress

*Figure 5.13* Strain and stress distributions of strengthened reinforced concrete section at working load.

with no load, simple modification of the equations presented in Section 5.5 for cracked section analysis of reinforced concrete is sufficient. Note the similarity between Figures 5.6 and 5.13.

The depth of the neutral axis can be computed using the equation:

$$\frac{b(kd)^2}{2} + (n-1)A_s'(kd - d') = nA_s\,(d - kd) + n_f A_f\,(h - kd) \qquad (5.47)$$

$$\text{where, } n_f = \frac{E_f}{E_c} \qquad (5.48)$$

$E_f$ = modulus of elasticity of composite fibers.

$A_f$ = area of fibers in the composite.

The cracked moment of inertia:

$$I_{cr} = \frac{b(kd)^3}{2} + nA_s(d-kd)^2 + (n-1)A_s'(kd-d')^2 + n_f A_f\,(h-kd)^2 \quad (5.49)$$

The maximum stress for a given moment, $M$, in concrete, $f_c$, steel, $(f_s - f_s')$, and composite, $f_f$, can be estimated using:

$$f_c = \frac{M}{I_{cr}}kd \qquad (5.50)$$

$$f_s = \frac{M}{I_{cr}}(d - kd)n \qquad (5.51)$$

$$f_s' = \frac{M}{I_{cr}}(kd - d')n \qquad (5.52)$$

$$f_f = \frac{M}{I_{cr}}(h - kd)n_f \qquad (5.53)$$

However, since the composites are installed when loads are present, the computations have to be done in two stages. This can be explained using the moment-curvature relationship shown in Figure 5.14. For any given moment, $M$, the curvature could range from a maximum of $\psi_1$ for the unstrengthened beam to a minimum of $\psi_2$ for the beam strengthened at zero load. The actual curvature, $\psi_3$, will be between $\psi_1$ and $\psi_2$, and the magnitude of the difference will depend on the load level at which the strengthening was applied. The stress will have similar variation. Even though the differences look significant in Figure 5.14, the numerical figures will not be significant because of the low strain values. The stresses and the equivalent stiffnesses for computing deflection can be estimated using the following procedure.

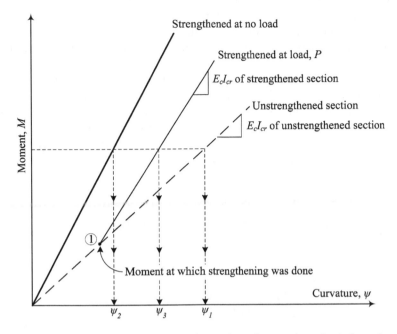

*Figure 5.14* Moment-curvature relationship of strengthened reinforced concrete beam.

*Step 1*: Compute the moment at which the strengthening was applied, $M_r$.

*Step 2*: Calculate $I_{cr}$, $I_e$, and stresses using the equations presented in Sections 5.5 and 5.6 for the moment in Step 1.

*Step 3*: Compute $I_{cr}$, $I_e$, and stresses using equations presented in this section for $(M - M_r)$. Here, $M$ is the moment for which stresses and deflections are needed.

*Step 4*: Add all the stresses and deflections of Steps 2 and 3.

Stresses and deflections can be computed by using the values of strengthened section for $M$. However, the estimated values will be lower than the actual values. Again, the differences may not be significant.

## 5.16  Allowable stresses check for upgraded loads

The allowable stresses at working loads are as follows:

Concrete (compression) $0.45f_c'$
Mild steel (tension) $0.8f_y$
Mild steel (compression) $0.4f_y$
Carbon composite (fibers, tension) $0.33f_u$
Glass composite (fibers, tension) $0.33f_u$

It is suggested to use a reduction factor between 0.9 and 1.0 for carbon and between 0.75 and 1.00 for glass in cases of severe environmental exposure. For sustained loading, the glass fiber stress should be further reduced using a factor of 0.30 to prevent creep fracture (ACI Committee 440, 2002).

The equations presented in Section 5.15 can be combined with the aforementioned stresses to check for allowable stresses. This check is done after estimating the required composite area using strength design procedures. Therefore, all the geometric details, area of composite, upgraded loads and load levels at which strengthening was done are known. The sequence of computations is presented in Figure 5.15.

## 5.17   Computation of maximum stresses and deflections for strengthened beams

### Example 5.10: Strengthened rectangular beam

Using the details of Examples 5.1 and 5.3, compute the stresses and deflection if the section was strengthened using three layers of carbon sheet. Equivalent thickness of carbon sheet was 0.0043 in. The sheet was applied over a width of 6.6 in. The modulus of elasticity for fibers, $E_f$, was $33 \times 10^6$ psi and the fracture stress, $f_{fu}$, was 550,000 psi. The fiber sheets (composite) were installed on the unstressed beam. The total maximum service load is 12,000 lb. This service load represents 60% of the experimental ultimate load (M'Bazaa et al., 1996).

Solution:

Since the composite was installed on the unstressed beam, only the second part of the calculations, Figure 5.15(b), is needed. The cracking moment and $I_g$ were taken from Example 5.1.

Given:

$b = 8$ in.

$d = 12$ in.

$h = 12$ in.

$f_y = 60,000$ psi

$f_{fu} = 550,000$ psi

$E_f = 33 \times 10^6$ psi

$f_c' = 5500$ psi

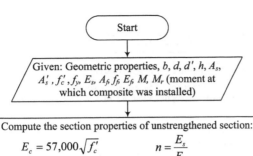

Start

Given: Geometric properties, $b$, $d$, $d'$, $h$, $A_s$, $A_s'$, $f_c'$, $f_y$, $E_s$, $A_f$, $f_f$, $E_f$, $M$, $M_r$ (moment at which composite was installed)

Compute the section properties of unstrengthened section:

$$E_c = 57,000\sqrt{f_c'} \qquad\qquad n = \frac{E_s}{E_c}$$

Solve for $kd$ and cracked moment of inertia:

Rectangular section or flanged section when $kd \le h_f$:

$$\frac{b(kd)^2}{2} + (n-1)A_s'(kd - d') = nA_s(d - kd)$$

$$I_{cr} = \frac{b(kd)^3}{3} + nA_s(d - kd)^2 + (n-1)A_s'(kd - d')^2$$

Flanged section when $kd > h_f$:

$$\frac{b(kd)^2}{2} - \left(\frac{b - b_w}{2}\right)(kd - h_f)^2 + (n-1)A_s'(kd - d') = nA_s(d - kd)$$

$$I_{cr} = \frac{b(kd)^3}{3} - \frac{(b - b_w)}{3}(kd - h_f)^3 + nA_s(d - kd)^2 + (n-1)A_s'(kd - d')^2$$

Compute stresses at $M_r$:

$$f_{c1} = \frac{M_r}{I_{cr}}kd \qquad f_{s1} = \frac{M_r}{I_{cr}}(d - kd)n \qquad f_{s1}' = \frac{M_r}{I_{cr}}(kd - d')n$$

Compute deflection at $M_r$:

$f_r = 7.5\sqrt{f_c'}$ for normal weight concrete

$I_g = \dfrac{bh^3}{12}$ for rectangular section

$\bar{y} = \dfrac{h}{2}$ for rectangular section

For T-section, use eqns 5.9 to 5.12

$$M_{cr} = \frac{I_g}{(h - \bar{y})}f_r \qquad I_e = I_{cr} + (I_g - I_{cr})\left(\frac{M_{cr}}{M_r}\right)^3$$

$$\delta_1 = \frac{P_r L^n}{kE_c I_e} \qquad n \text{ and } k \text{ depend on load and support conditions}$$

Continue for strengthened section

(a) Unstrengthened section

*Figure 5.15* Flowchart for checking allowable stresses.

Compute the section properties of strengthened section:

It is assumed that the composite is placed at a distance of $h$ from extreme compression fiber. If the composite is placed at any other location, use the appropriate value instead of $h$. This is also true for computation of $I_{cr}$ and stresses.

Solve for new $kd$ and cracked moment of inertia:
Rectangular section or flanged section when $kd \leq h_f$:

$$\frac{b(kd)^2}{2} + (n-1)A_s'(kd-d') = nA_s(d-kd) + n_f A_f(h-kd) \qquad n_f = \frac{E_f}{E_c}$$

$$I_{cr} = \frac{b(kd)^3}{3} + nA_s(d-kd)^2 + (n-1)A_s'(kd-d')^2 + n_f A_f(h-kd)^2$$

Flanged section when $kd > h_f$:

$$\frac{b(kd)^2}{2} - \left(\frac{b-b_w}{2}\right)(kd-h_f)^2 + A_s'(kd-d') = nA_s(d-kd) + n_f A_f(h-kd)$$

$$I_{cr} = \frac{b(kd)^3}{3} - \frac{(b-b_w)}{3}(kd-h_f)^3 + nA_s(d-kd)^2 + (n-1)A_s'(kd-d')^2 + n_f A_f(h-kd)^2$$

Compute stresses

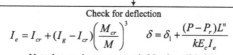

$$f_c = f_{c1} + \frac{(M-M_r)}{I_{cr}}kd \qquad f_s = f_{s1} + \frac{(M-M_r)}{I_{cr}}(d-kd)n$$

$$f_s' = f_{s1}' + \frac{(M-M_r)}{I_{cr}}(d-d')n \qquad f_f = \frac{(M-M_r)}{I_{cr}}(h-kd)n_f$$

Check for allowable stresses

$$f_c \leq 0.45 f_c', \quad f_s \leq 0.80 f_y, \quad f_s' \leq 0.40 f_y$$

$f_f \leq (0.90)(0.33) f_{fu}$ for carbon

$f_f \leq 0.75(0.33)(0.30) f_{fu}$ for glass subjected to sustained loads

Check for deflection

$$I_e = I_{cr} + (I_g - I_{cr})\left(\frac{M_{cr}}{M}\right)^3 \qquad \delta = \delta_1 + \frac{(P-P_r)L^n}{kE_c I_e}$$

Note that maximum moment is $M$ and not $(M-M_r)$

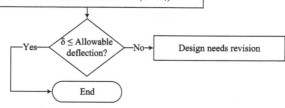

—Yes— $\delta \leq$ Allowable deflection? —No→ Design needs revision

End

**(b) Strengthened section**

*Figure 5.15* (Continued).

*Note that there are two sets of values for $kd$, $I_{cr}$, and $I_e$.
If any of the conditions is violated, the design needs revision.

From the solution of Example 5.1:

$$E_c = 4.2 \times 10^6 \, \text{psi}$$
$$n = 29/4.2 = 6.9$$
$$I_g = 1440 \, \text{in.}^4$$
$$M_{cr} = 133{,}440 \, \text{in.-lb}$$
$$A_s = 0.33 \, \text{in.}^2$$

For composite:

$$n_f = \frac{33 \times 10^6}{4.2 \times 10^6} = 7.9$$
$$A_f = 3 \times 6.6 \times 0.0043 = 0.085 \, \text{in.}^2$$

Computation of depth of neutral axis:

$$\frac{b(kd)^2}{2} = nA_s \left(d - kd\right) + n_f A_f \left(h - kd\right)$$
$$\frac{8(kd)^2}{2} = 6.9 \times 0.33 \left(10 - kd\right) + 7.9 \times 0.085 \left(12 - kd\right)$$
$$\text{or } (kd)^2 + 0.7(kd) - 7.7 = 0$$
$$kd = 2.5 \, \text{in.}$$
$$I_{cr} = \frac{8 \times 2.5^3}{3} + 6.0 \times 0.33 \left(10 - 2.5\right)^2 = 230 \, \text{in.}^4$$
$$P = 12{,}000 \, \text{lb}$$

$$\text{Moment, } M_u = \frac{12{,}000}{2} \times 40 = 240{,}000 \, \text{in.-lb}$$

$$f_c = \frac{240{,}000}{230} \times 2.5 = 2609 = 0.47 f_c'$$
$$f_c > 0.45 f_c'$$
$$f_s = \frac{240{,}000}{230} \times (10 - 2.5) \times 6.9 = 54{,}000 = 0.9 f_y$$
$$f_s > 0.8 f_y$$

$$f_f = \frac{240,000}{230} \times (12 - 2.5) \times 7.9 = 78,313 = 0.14 f_{fu}$$

$$f_f < 0.9 \times 0.33 f_{fu}$$

Both the concrete stress and the steel stress exceed the allowable stresses. This type of condition could occur if composite is used to increase the strength of the beam excessively.

If an allowable concrete stress of $0.45 f_c' = 0.45 \times 5500 = 2475$ is used, then allowable moment, $\dfrac{2475 \times 230}{2.5} = 227,700$ in.-lb.

If an allowable stress of $0.8 f_y$ or 48,000 psi is used for steel, then allowable moment, $\dfrac{48,000 \times 230}{(10 - 2.5) \times 6.9} = 213,333$ in.-lb.

Therefore, for the strengthened beam, the maximum allowable working load moment is about 210,000 in.-lb. For this moment, stresses in concrete and composite will be within the allowable stress limits.

Computation of deflection:

The deflection is computed for the maximum moment of 210,000 in.-lb. For this moment, the load is:

$$P = \frac{210,000}{40} \times 210,500 \text{ lb}$$

$$I_e = 230 + (1440 - 230)\left(\frac{133,440}{210,000}\right)^3 = 540 \text{ in.}^4$$

$$\delta = \frac{10,500(40)(3 \times 120^2 - 4 \times 40^2)}{48 \times 4.2 \times 10^6 \times 540} = 0.142 \text{ in.}$$

$$\delta = \frac{L}{845}$$

The deflection will satisfy the requirements for most conditions. Note that the experimental deflection for this load is about 0.17 in. (M'Bazaa *et al.*, 1996).

### Example 5.11: Strengthened T-beam

Compute the stresses and deflection for the T-beam of Example 5.2. The composite was installed at load of 15,000 lb. The composite consisted of 3–5-in. wide layers of carbon composite. The equivalent thickness was 0.0043 in. and composite modulus was $33 \times 10^6$ psi. Fracture stress, $f_{fu} = 550,000$ psi. Compute the stresses and deflection for a load of 42,000 lb.

Solution:

Given:

$$f'_c = 4500\,\text{psi}$$

$$b = 15\,\text{in.}$$

$$b_w = 6\,\text{in.}$$

$$d = 14.5\,\text{in.}$$

$$h = 16\,\text{in.}$$

$$h_f = 4\,\text{in.}$$

$$A_s = 2.0\,\text{in.}^2$$

$$E_s = 29 \times 10^6\,\text{psi}$$

$$E_c = 3.8 \times 10^6\,\text{psi}$$

$$n = 7.6$$

$$M_{cr} = 159,214\,\text{in.-lb}$$

$$I_g = 3039\,\text{in.}^4$$

$$A_f = 3 \times 5 \times 0.0043 = 0.065\,\text{in.}^2$$

Computation of stresses and deflection for unstrengthened beam:
Load when composite was installed, $P_r = 15,000\,\text{lb}$.

$$M_r = \frac{P}{2} \times 42 = \frac{15,000}{2} \times 42 = 315,000\,\text{in.-lb}$$

For unstrengthened beam:

$$I_{cr} = 1976\,\text{in.}^4 \text{ (from Example 5.4)}$$

$$kd = 4.52\,\text{in.}$$

$$\text{Stresses, } (f_c)_r = \frac{M_r}{I_{cr}}kd = \frac{315,000}{1976} \times 4.52 = 721\,\text{psi}$$

$$(f_s)_r = \frac{315,000}{1976}(14.5 - 4.52)\,7.6 = 12,091\,\text{psi}$$

$$I_e = 1976 + (3039 - 1976)\left(\frac{159,214}{315,000}\right)^3 = 2113\,\text{in.}^4$$

$$\delta = \frac{15,000\,(42)(3 \times 92^2 - 4 \times 42^2)}{48 \times 3.8 \times 10^6 \times 2113} = 0.04\,\text{in.}$$

For strengthened beam:

$$n_f = \frac{33 \times 10^6}{3.8 \times 10^6} = 8.7$$

Since the depth of the neutral axis was below flange thickness for the unstrengthened beam, it should be larger than the flange thickness beam.

$$\frac{15(kd)^2}{2} - (15 - 6)\frac{(kd - 4)^2}{2} = 7.6 \times 2\left(14.5 - kd\right) + 8.7$$

$$\times 0.065\left(16 - kd\right)$$

$kd = 4.6\,\text{in.}$

$$I_{cr} = \frac{15 \times 4.6^3}{3} - (15 - 6)\frac{(4.6 - 4)^3}{3} + 7.6 \times 2\left(14.5 - 4.6\right)^2$$

$$+ 8.7 \times 0.065\left(16 - 4.6\right)^2$$

$$= 2049\,\text{in.}^4$$

$$(M - M_r) = (42{,}000 - 15{,}000) \times \frac{42}{2} = 567{,}000\,\text{in.-lb}$$

Computation of stresses for $(M - M_r)$:

$$f_c = 721 + \frac{567{,}000}{2049} \times 4.6$$

$$= 1994\,\text{psi} = 0.44 f_c'$$

$f_c < 0.45 f_c'$

$$f_s = 12{,}091 + \frac{567{,}000}{2049} \times (14.5 - 4.6) \times 7.6 = 32{,}911\,\text{psi} = 0.55 f_y$$

$f_s < 0.8 f_y$

$$f_f = \frac{567{,}000}{2049} \times (16 - 4.6) \times 8.7 = 32{,}260\,\text{psi} = 0.06 f_{fu}$$

$f_f < 0.9 \times 0.33 f_{fu}$

Computation of deflection:

$$I_e = 2049 + (3039 - 2049)\left(\frac{159{,}214}{882{,}000}\right)^3 = 2055 \text{ in.}^4$$

$$\delta = 0.04 + \frac{(42{,}000 - 15{,}000 \times 42(3 \times 92^2 - 4 \times 42^2)}{48 \times 3.8 \times 10^6 \times 2055} = 0.10 \text{ in.}$$

$$\delta = \frac{L}{920}$$

## 5.18 Strength analysis – nominal resisting moment, $M_n$: strengthened reinforced concrete

The nominal moment capacity of the strengthened cross-section is of great importance because the upgraded loads will be estimated using this value. The basic principles of analysis are the same as the principles used for reinforced concrete. The tension force provided by the composite becomes an additional contribution. Typical strain and stress distributions for a rectangular section are shown in Figure 5.16.

The depth of the neutral axis is controlled by the force equilibrium. At failure, if failure occurs by crushing of concrete, the strain and stress for concrete are assumed as 0.003 and $0.85f_c'$, respectively. Therefore, the compressive force capacity of concrete:

$$C_c = 0.85f_c'ba \tag{5.54}$$

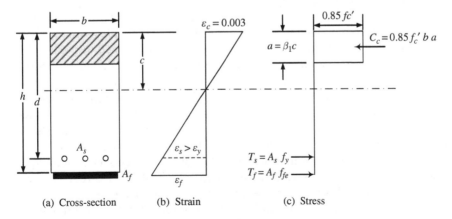

(a) Cross-section     (b) Strain     (c) Stress

*Figure 5.16* Typical strain and stress distribution of a strengthened section at failure.

This failure mechanism and force contribution is well established (ACI Committee 318, 2005). The force contribution of steel is:

$$T_s = A_s f_y \qquad (5.55)$$

Since the original beams were designed as under-reinforced sections, this force contribution is also well established. At larger strains (>0.008 for grade 60 steel), strain-hardening may occur. However, this extra force contribution is neglected to obtain a conservative estimate of moment capacity.

Unstrengthened beams fail by steel yielding followed by crushing of concrete. The additional composite force at failure:

$$T_f = A_f f_{fe} \qquad (5.56)$$

This additional force leads to the following possible failure modes:

- crushing of concrete;
- fracture of composite;
- balanced failure in which both concrete and composite fail simultaneously;
- shear/tension delamination of concrete cover or debonding of composite from concrete substrate.

For composites, the fracture strains are measured using tension (coupon) tests. These values can be obtained from the manufacturer's technical data sheets. For example, T300 carbon fibers fracture at a strain of 0.015. If it is assumed that this fracture strain can be achieved without encountering other failure mechanisms such as delamination of composite plates, the failure conditions can be set up easily.

However, a number of failure mechanisms, other than fracture of composite, can occur as explained in Section 5.17. The most prevalent among these are failure due to delamination. Models have been developed to predict this failure mechanism (Arduini *et al.*, 1997; Kurtz, 2000; Malek *et al.*, 1998). Due to the complexity of the models, the following procedure, which is based on ACI Committee 440 (2002), is recommended for computing the composite strain at failure, $\varepsilon_{fe}$. For the purpose of these calculations, it is recommended to use only the fiber properties. Since the matrix used in the composite has a very low modulus, its effect is neglected. Therefore, if composite thickness, modulus, and fracture strains are given, they should be converted into equivalent fiber area, modulus, and fracture strain. Typically, fracture strains will be the same; modulus will increase, and the area will decrease.

The manufacturer's recommended (or guaranteed) fracture strain of composite (fiber), $\varepsilon_{fu}$, should be multiplied by a factor, $\kappa_m$, to account for possible debonding or delamination. This reduction factor is based on the hypothesis that laminates with greater stiffness are more prone to delamination. Therefore, the factor depends on unit stiffness defined by $nE_f t_f$, where $n$ is the number of layers and $t_f$ is the thickness. Note that $nt_f$ is the area of cross-section of fiber per unit width. Therefore, if carbon plates are used instead of sheets, replace $nt_f$ with the thickness of the plate. Again, use the equivalent fiber thickness and fiber modulus instead of composite area and composite modulus. The factor, $\kappa_m$, also is limited to a maximum of 0.9.

$$\kappa_m = \left\{ 1 - \frac{nE_f t_f}{2,400,000} \quad \text{if } nE_f t_f \leq 1,200,000 \right.$$

$$\kappa_m = \left\{ \frac{600,000}{nE_f t_f} \quad \text{if } nE_f t_f > 1,200,000 \right. \tag{5.57}$$

If the manufacturer's recommended fracture strain is $\varepsilon_{fu}$, the usable fracture strain at failure is:

$$\varepsilon_{fe} = \kappa_m \varepsilon_{fu} \tag{5.58}$$

Once $\varepsilon_{fe}$ is established, the tension force contribution of the composite can be taken as $A_f E_f \varepsilon_{fe}$, if failure occurs by fracture of composite. If failure occurs by fracture of composite or simultaneous fracture of composite and crushing of concrete, the force equilibrium equation at failure can be written as:

$$0.85 f'_c ba = A_s f_y + A_f E_f \varepsilon_{fe} \tag{5.59}$$

It is assumed that even though failure is by fracture of composite, concrete undergoes sufficient strain and the rectangular stress block assumption for concrete is still valid. However, if a more accurate evaluation is needed, the following equations can be used to compute the force contribution of concrete and the depth of the neutral axis (Todeschini *et al.*, 1964).
Force contribution of concrete:

$$C = 0.85 f'_c ba \tag{5.60}$$

$$\text{where, } a = \beta_1 c \tag{5.61}$$

$$\beta_1 = 2 - \frac{4\left[\left(\frac{\varepsilon_c}{\varepsilon'_c}\right) - \tan^{-1}\left(\frac{\varepsilon_c}{\varepsilon'_c}\right)\right]}{\left[\left(\frac{\varepsilon_c}{\varepsilon'_c}\right) \ln\left(1 + \left(\frac{\varepsilon_c^2}{\varepsilon_c'^2}\right)\right)\right]} \tag{5.62}$$

$$c = \frac{A_s f_y + A_f E_f \varepsilon_{fu}}{\gamma f'_c \beta_1 b} \tag{5.63}$$

$$\gamma = \frac{0.9 \ln \left(1 + \left(\frac{\varepsilon_c^2}{\varepsilon_c'^2}\right)\right)}{\beta_1 \left(\frac{\varepsilon_c}{\varepsilon_c'}\right)} \tag{5.64}$$

where $\varepsilon_c' = 1.71 f_c'/E_c$ and $\tan^{-1}(\varepsilon_c/\varepsilon_c')$ is computed in radians.

If failure occurs by crushing of concrete first, then strain in the composite has to be computed using force and strain compatibility equations.

For strain compatibility:

$$\frac{0.003}{\varepsilon_f} = \frac{c}{h - c} \tag{5.65}$$

For force equilibrium:

$$0.85 f_c' b \beta_1 c = A_s f_y + A_f E_f 0.003 \left(\frac{h - c}{c}\right) \tag{5.66}$$

Equation (5.66) can be solved for $c$ and:

$$a = \beta_1 c \tag{5.67}$$

$$\varepsilon_c' = 0.003 \left(\frac{h - c}{c}\right) \tag{5.68}$$

Once the depth of the neutral axis is established, the nominal moment capacity, $M_n$, can be estimated using the moment equilibrium equation:

$$M_n = A_s f_y \left(d - \frac{a}{2}\right) + A_f E_f \varepsilon_f (\text{or } \varepsilon_{fe}) \left(h - \frac{a}{2}\right) \tag{5.69}$$

If the composite plate is at a location other than the extreme tension face, the corresponding value instead of $h$ should be used. For beams with compression reinforcement and flanged sections, equations similar to (5.59), (5.66), and (5.69) can be developed based on force and moment equilibrium. The principles are the same for unstrengthened and strengthened beams.

For beams with compression reinforcement, $A_s'$, compute the depth of the stress block using the equations below. If failure occurs by fracture of composite or by simultaneous crushing of concrete and fracture of composite:

$$0.85 f_c' b a + A_s' f_s' = A_s f_y + A_f E_f \varepsilon_{fe} \tag{5.70}$$

If failure is by crushing of concrete first:

$$0.85f_c'b\beta_1 c + A_s'f_s' = A_sf_y + A_fE_f0.003\left(\frac{h-c}{c}\right) \tag{5.71}$$

$$a = \beta_1 c \tag{5.72}$$

Nominal moment:

$$M_n = A_sf_y\left(d - c\right) + A_fE_f\varepsilon_f(\text{or } \varepsilon_{fe})\left(h - c\right) + A_s'f_s'\left(c - d'\right)$$
$$+ 0.85f_c'ba\left(c - a/2\right) \tag{5.73}$$

The stress, $f_s'$, has to be computed using trial-and-adjustment, Section 5.11. If the compression steel is close to the depth of the stress block, its contribution can be neglected.

For flanged sections, if $c \le h_f$, treat it as a rectangular beam with $b$, $d$, and $h$.

If $c > h_f$ and failure is by fracture of composite or balanced failure:

$$0.85f_c'b_wa + 0.85f_c'\left(b - b_w\right)h_f = A_sf_v + A_fE_f\varepsilon_{fe} \tag{5.74}$$

For failure by crushing of concrete first:

$$0.85f_c'b_wc/\beta_1 + 0.85f_c'\left(b - b_w\right)h_f = A_sf_y + A_fE_f\left(\frac{h-c}{c}\right) \tag{5.75}$$

$$a = \beta_1 c$$

$$M_u = 0.85f_c'\left(b - b_w\right)h_f\left(c - \frac{h_f}{2}\right) + 0.85f_c'b_wa\left(c - \frac{a}{2}\right)$$
$$+ A_sf_y\left(d - c\right) + A_fE_f\varepsilon_f(\text{or } \varepsilon_{fe})\left(h - c\right) \tag{5.76}$$

For the entire computation, strain in the composite should be adjusted for the load present when the composite was applied. The sequence of calculations for rectangular and flanged sections is presented in flowcharts, Figures 5.17 and 5.18. Width of flange as well as minimum and maximum reinforcement should be checked for unstrengthened beams. For sections with compression reinforcement, add the contribution to the compression side. Fracture strain for the composite should be based on the discussion presented in the beginning of the section. The following are the major steps:

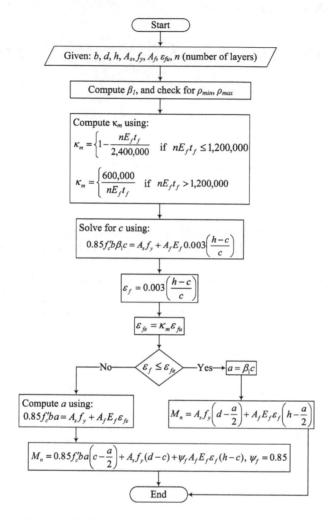

*Figure 5.17* Flowchart for computation of $M_n$: strengthened rectangular section.

*Step 1*: Assuming failure occurs by crushing of concrete first, compute $c$ and strain in composite at failure, $\varepsilon_f$.

*Step 2*: If $\varepsilon_f \leq \varepsilon_{fe}$, proceed to compute $M_n$.

If $\varepsilon_f \geq \varepsilon_{fe}$, compute $a$ and $M_n$ based on failure by fracture of composite.

### 5.18.1 Reduction factor for contribution of composite nominal strength

Even though composites have been used for more than three decades, their widespread use as a reinforcement is relatively new. In order to account for

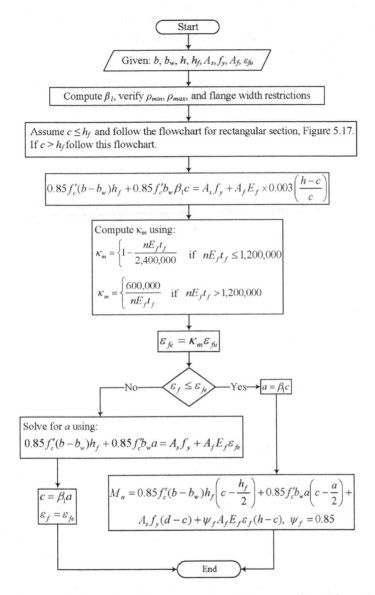

*Figure 5.18* Flowchart for computation of $M_n$: strengthened flanged section.

any uncertainty, ACI committee 440 on FRP (2002) recommends a reduction factor of 0.85. This reduction factor, $\psi_f$, is applied only to the contribution of the composite.

The moment is computed by multiplying the total compressive force or the total tensile force with lever arm. Typically, the moment is computed about the resultant location of the compression force. However, since the reduction factor is applied only on the tension force of the composite, the moment should be computed about the neutral axis. This procedure will ensure that a consistent reduction factor is applied to the composite force.

The aforementioned discussion leads to the modification of Equations (5.69), (5.73), and (5.76) for rectangular section with tension reinforcement, tension and compression reinforcement, and flanged section, respectively. The corresponding new equations are:

$$M_n = 0.85f_c'ba\left(c - a/2\right) + A_sf_y\left(d - c\right) + \psi_f A_f E_f \varepsilon_f (\text{or } \varepsilon_{fe})\left(h - c\right)$$

$$(5.69a)$$

$$M_n = 0.85f_c'ba\left(c - a/2\right) + A_sf_y\left(d - c\right) + A_s'f_s'\left(c - d'\right)$$
$$+ \psi_f A_f E_f \varepsilon_f(\text{or } \varepsilon_{fe})\left(h - c\right)$$

$$(5.73a)$$

$$M_n = 0.85f_c'\left(b - b_w\right)h_f\left(c - \frac{h_f}{2}\right) + 0.85f_c'b_wa\left(c - \frac{a}{2}\right)$$
$$+ A_sf_y\left(d - c\right) + \psi_f A_f E_f \varepsilon_f(\text{or } \varepsilon_{fe})\left(h - c\right)$$

$$(5.76a)$$

These modifications are incorporated into the flowcharts, Figures 5.17 and 5.18.

### 5.18.2  Reduction factor for long-term exposure

The strength, modulus, and fracture strain reported by manufacturers should be assumed to be initial properties. In the actual structures, there is the possibility of some loss of strength due to long-term exposure. The reduction will depend on both fiber and matrix type and exposure conditions.

Based on ACI 440 (2002) recommendations, the reduction factors recommended for three common fiber types are presented in Table 5.1. The matrix is assumed an epoxy for all cases.

The strength calculations should be repeated for these allowable stresses (strains). The analysis for short-term load has to be done independently because the failure mechanisms could be different. For example, if the maximum allowable fracture strain is reduced, the failure mechanism could change from initial crushing of concrete to initial fracture of composite plate. In other words, the strengthened beam could fail by fracture of composite after 10 years of exposure to aggressive environmental conditions.

*Table 5.1* Environmental reduction factors (ACI Committee 440, 2002)

| Exposure conditions | Fiber type | Reduction factor, $C_E$ |
|---|---|---|
| Interior | Carbon | 0.95 |
|  | Aramid | 0.85 |
|  | Glass | 0.75 |
| Exterior (bridges, parking decks, and so on) | Carbon | 0.85 |
|  | Aramid | 0.75 |
|  | Glass | 0.65 |
| Aggressive environment (chemical plants, water treatment plants, etc.) | Carbon | 0.85 |
|  | Aramid | 0.70 |
|  | Glass | 0.50 |

## 5.19 Typical failure modes of composite, and estimation of composite strain at failure

In a perfect system, it should be possible to realize the full strength of the composite. For example, at failure, standard carbon composites should reach a strain of 0.015. However, as mentioned in the previous section, failure could occur before the fracture strain is reached, say, due to delamination. The possible failure patterns and methods are shown in Figure 5.19 and discussed in this section.

### 5.19.1 Sheets, fabrics, and tows

In the application process, composite sheets are attached to sound concrete. Typically, the concrete is primed before the application of the sheet. The weak link is the concrete adjacent to the adhesive. Delamination can occur due to:

1  interfacial shear failure Figure 5.19(a);
2  beam shear, Figure 5.19(b);
3  cover tension, Figure 5.19(c);
4  planar surface irregularities, Figure 5.19(d);
5  poor surface preparation.

Poor surface preparation should be avoided. The other mechanisms are briefly described in the following sections.

#### 5.19.1.1 Interfacial shear

The tensile force resisted by the composite plate is transferred to the beam by interfacial shear (Figure 5.19(a)). When this shear stress exceeds the shear strength of the interface, debonding occurs. As mentioned earlier, this failure

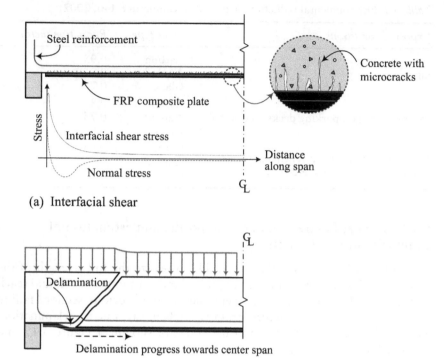

(a) Interfacial shear

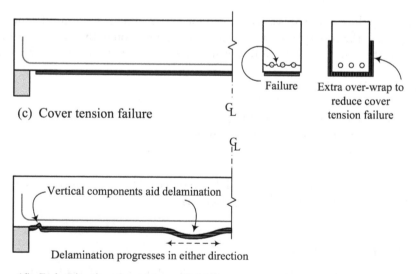

Delamination progress towards center span

(b) Delamination caused by excessive shear deformation

(c) Cover tension failure

Failure    Extra over-wrap to
reduce cover
tension failure

Vertical components aid delamination

Delamination progresses in either direction

(d) Delamination due to imperfections

*Figure 5.19* Delamination failure mechanisms (Source: ACI Committee 440, 2002).

occurs in the concrete just above the adhesive. Typical shear strength of concrete, about 800 psi, is high enough to transfer composite forces. However, as the beam bends, microcracks occur in the concrete just above the adhesive (epoxy) to compensate for the strains in the composite. Note that tensile strain capacity of concrete is only about 0.00025, and the composite sheets can sustain more than 0.015. Most epoxies used for adhesive can also sustain large strains and do not crack. Hence, a layer of concrete with microcracks becomes very susceptible for interlaminar shear failure. Theories are being developed for the prediction of this important failure mechanism (Arduini *et al.*, 1997; Kurtz, 2000; Malek *et al.*, 1998). In addition to delamination due to shear failure, peeling can occur at the termination point of the plate due to highly concentrated peeling stresses (Figure 5.19(a)). Even though the mechanism can be predicted with reasonable accuracy, it is recommended to use a lower strain for the composite at failure rather than go through extensive analysis. Based on extensive experimental evidence, it is recommended to use a fracture strain of 0.008 for carbon composites. For other composites, the manufacturer's recommendation can be followed. Note that, at a strain of 0.008, the section undergoes extensive curvature and stresses in the range of 200 ksi are generated.

In special situations where short bend lengths or thick plates are encountered, detailed analysis should be conducted. Kurtz's work (2000) provides detailed information on all three models and sample calculations. It is also advisable to stagger the termination of plies to avoid excessive stresses. A typical pattern recommended by ACI (2000) is shown in Figure 5.20.

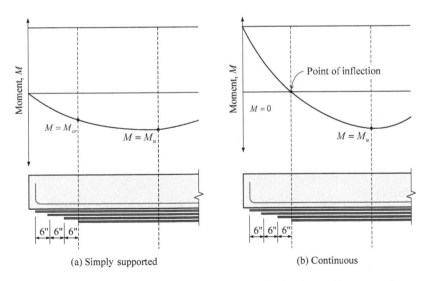

(a) Simply supported          (b) Continuous

*Figure 5.20* Recommended termination for multiple-ply applications (Source: ACI Committee 440, 2002).

### 5.19.1.2 Beam shear

When beams are strengthened for flexure, the shear capacity of the beams should also be increased. If excessive shear deformations occur before flexural failure, the large diagonal crack at the critical shear plane could push the plate to initiate the delamination (Figure 5.19(b)). The delamination will propagate toward the support first before continuing toward the center span. This failure can be avoided by ensuring adequate shear capacity.

### 5.19.1.3 Cover tension

The plane across the tension reinforcement becomes weak due to reduction in concrete area (Figure 5.19(c)). When this plane has to transfer the extra forces generated by the added composite plate, it could develop extensive cracking. In extreme cases, the concrete between the reinforcement and the bottom composite plate was found to crumble. Note that the concrete between steel (tension) reinforcement and the extreme tension develops flexural cracks at service loads and, hence, becomes weaker. For beams with heavy tension reinforcement placed at minimum spacing, it is advisable to provide over-wraps or U-wraps. These composite wraps confine the concrete, providing extra tension force along the thickness of the beam. The placement of over-wraps becomes a necessity for beams with bundled reinforcement.

### 5.19.1.4 Surface irregularities

If irregularities are present, they become the delamination initiators (Figure 5.19(d)). Both protrusions and low spots become critical locations. The surface should be ground to a smoothness recommended by manufacturers. Typically, the tolerance is less than one sixteenth of an inch or 1 mm.

## 5.20 Computation of nominal moment capacities: strengthened beams

*Example 5.12: Rectangular beam – composite applied at unstressed state*
Compute the nominal moment capacity using the details of Example 5.10. The composite was applied at unstressed state. (a) Assume a fracture strain of 0.008. for carbon. (b) Use the manufacturer's reported strain of 0.017.

Solution:

(a) Assume a fracture strain of 0.008 for carbon.

Given:

$b = 8$ in.

$d = 10$ in.

$b = 12\,\text{in.}$

$A_s = 0.33\,\text{in.}^2, E_s = 29 \times 10^6\,\text{psi}$

$A_f = 0.085\,\text{in.}^2, E_f = 33 \times 10^6\,\text{psi}$

$f_c' = 5500\,\text{psi}, f_y = 60{,}000\,\text{psi}$

The minimum and maximum reinforcement ratios have been checked (Example 5.7).

$\beta_1 = 0.775$

$$0.85 \times 5500 \times 8 \times 0.775c = 0.33 \times 60{,}000 + 0.085 \times 33$$

$$\times 10^6 \times 0.003 \left( \frac{12 - c}{c} \right)$$

$c^2 - 0.4c - 3.5 = 0$

$c = 2.1\,\text{in.}$

$$\varepsilon_f = \frac{12 - 2.1}{2.1} \times 0.003 = 0.0140 > \varepsilon_{fu} = 0.008$$

Therefore assume failure strain of 0.008 for composite.

$$0.85 \times 5500 \times 8 \times a = 0.33 \times 60{,}000 + 0.085 \times 0.008 \times 33 \times 10^6$$

$a = 1.13\,\text{in.}$

Nominal moment capacity:

$$M_n = 0.33 \times 60{,}000 \left( 10 - \frac{1.13}{2} \right) + 0.085 + 0.008 \times 33 \times 10^6$$

$$\left( 12 - \frac{1.13}{2} \right) = 443{,}401\,\text{in.-lb}$$

Note that if another failure mechanism (say, delamination) is not encountered, the strain in composite could reach 0.014 when concrete starts crushing. At this stage, assumption of rectangular stress block for concrete stress distribution is accurate. However, at a composite strain 0.008, the concrete did not reach the crushing stage. Rectangular block was assumed for simplicity. The following is the rigorous nonlinear analysis; refer to Equations (5.62)–(5.64).

Assume a maximum strain in concrete $\varepsilon_c = 0.002$:

$$\varepsilon_c' = \frac{1.7 \times 5500}{4.2 \times 10^6} = 0.00223$$

$$\beta_1 = 2 - \frac{4\left[\left(\frac{\varepsilon_c}{\varepsilon_c'}\right) - \tan^{-1}\left(\frac{\varepsilon_c}{\varepsilon_c'}\right)\right]}{\left[\left(\frac{\varepsilon_c}{\varepsilon_c'}\right)\ln\left(1 + \left(\frac{\varepsilon_c^2}{\varepsilon_c'^2}\right)\right)\right]} = 0.75$$

$$\gamma = \frac{0.9\ln\left(1 + \left(\frac{\varepsilon_c^2}{\varepsilon_c'^2}\right)\right)}{\beta_1\left(\frac{\varepsilon_c}{\varepsilon_c'}\right)} = 0.79$$

$$c = \frac{A_s f_y + A_f E_f \varepsilon_{fu}}{\gamma f_c' \beta_1 b}$$

$$= \frac{0.33 \times 60{,}000 + 0.085 \times 0.008 \times 33 \times 10^6}{0.79 \times 5500 \times 0.75 \times 8} = 1.62 \text{ in.}$$

Based on $c = 1.62$ in. and composite strain of 0.008 (Figure 5.21):

$$\varepsilon_c = 0.008\left(\frac{1.62}{10.38}\right) = 0.0012$$

$\varepsilon_c = 0.0012 <<$ assumed strain of 0.002

For a strain, $\varepsilon_c = 0.0012$

$\beta_1 = 0.71$

$\gamma = 0.6$

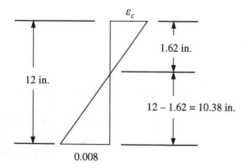

*Figure 5.21* Strain distribution for beam in Example 5.12.

$c = 2.25$ in.

$$\varepsilon_c = 0.008 \times \frac{2.25}{9.75} = 0.0018$$

For a strain, $\varepsilon_c = 0.0016$

$\beta_1 = 0.72$

$\gamma = 0.72$

$c = 1.85$

$\varepsilon_c = 0.0015$

For a strain, $\varepsilon_c = 0.00155$

$\beta_1 = 0.718$

$\gamma = 0.71$

$c = 1.87$ in.

$\varepsilon_c = 0.00155$

$a = 0.718 \times 1.87 = 1.35$ in.

(compared with 1.13 in. obtained using standard rectangular block)

Nominal moment,

$$M_n = 0.33 \times 60,000 \left(10 - \frac{1.35}{2}\right) + 0.085 \times 0.008 \times 33$$

$$\times 10^6 \left(12 - \frac{1.35}{2}\right)$$

$$= 438,786 \text{ in.-lb}$$
(approximate: 493,401 in.-lb)

Error based on approximate analysis = 1.06%.

The reduction factor, $\psi_f$, is not used for computation of $M_n$ so that the values can be compared with experimental results.

(b) Use the manufacturer's reported strain of 0.017.

Given:

$\varepsilon_{fe} = \kappa_m \times 0.017$

$nt_f E_f = 3 \times 0.0043 \times 33 \times 10^6 = 425,700$

Therefore

$$\kappa_m = 1 - \frac{425,700}{2,400,000} = 0.823$$

$\varepsilon_{fe} = 0.014$

This failure strain results in balanced failure.

$$0.85 \times 5500 \times 8 \times a = 0.33 \times 60{,}000 + 0.085 \times 0.014 \times 33 \times 10^6$$

$$a = 1.6 \, \text{in.}$$

$$c = 2.07 \, \text{in.}$$

$$M_n = 0.85 \times 5500 \times 8 \times 1.6 \left(2.07 - \frac{1.6}{2}\right) + 0.33 \times 60{,}000$$

$$(10 - 2.07) + 0.85 \times 0.085 \times 0.014 \times 33 \times 10^6 \, (12 - 2.07)$$

$$= 564{,}469 \, \text{in.-lb}$$

The experimental maximum moment was 448,000 in.-lb and the carbon strain at failure was 0.008 (M'Bazaa *et al.*, 1996). The error in moment computation is 1% based on the strain of 0.008 for composite.

*Example 5.13: Nominal moment capacity of strengthened T-beam*
Compute the nominal moment capacity using the details of Example 5.11. Manufacturer's guaranteed fracture strain is 0.017.

Solution:

  Given:

$$f_c' = 4500 \, \text{in.-lb}$$

$$b = 15 \, \text{in.}$$

$$b_w = 6 \, \text{in.}$$

$$d = 14.5 \, \text{in.}$$

$$h = 16 \, \text{in.}$$

$$h_f = 4 \, \text{in.}$$

$$A_s = 2.0 \, \text{in.}^2$$

$$E_s = 29 \times 10^6 \, \text{psi}$$

$$A_f = 0.065 \, \text{in.}^2$$

$$E_f = 33 \times 10^6 \, \text{psi}$$

$$\beta_1 = 0.825$$

The reinforcement satisfies minimum and maximum requirements.

Assuming $c < h_f$:

$$0.85 \times 4500 \times 15 \times 0.825c = 2.0 \times 60,000 + 0.065 \times 0.003$$

$$\times 33 \times 10^6 \times 0.003 \left(\frac{16 - c}{c}\right)$$

$c = 3.1 < h_f = 4$

$$\varepsilon_f = \left(\frac{16 - 3.1}{3.1}\right) \times 0.003 = 0.0155$$

$$nt_f E_f = 3 \times 0.0043 \times 33 \times 10^6 = 425,700 \kappa_m = 1 - \frac{425,700}{2,400,000}$$

$$= 0.823$$

$\varepsilon_{fe} = 0.823 \times 0.017 = 0.014$

Therefore, failure is initiated by fracture of composite.

$\varepsilon_{fe} = 0.014$

$0.85 \times 4500 \times 15 \times a = 2.0 \times 60,000 + 0.065 \times 33 \times 10^6 \times 0.014$

$a = 2.62 < h_f$

$c = 3.18$ in.

Nominal moment:

$$M_n = 0.85 \times 4500 \times 15 \times 2.62 \left(3.18 - \frac{2.62}{2}\right) + 2.0$$

$$\times 60,000 \,(14.5 - 3.18) + 0.85 \times 0.065 \times 0.014 \times 33 \times 10^6$$

$$\times 0.014 \,(16 - 3.18)$$

$$= 1,966,740 \text{ in.-lb}$$

Note that maximum concrete strain at the initiation of failure is less than 0.003, and the assumption of rectangular stress block is approximate.

*Example 5.14: Rectangular beam – composite applied with load present*
Repeat Example 5.12 if the composite is placed when the load was 7000 lb, producing a maximum moment of 140,000 in.-lb. Assume a failure strain, $\varepsilon_f$, of 0.014 for the composite.

Solution:

Using cracked section properties:

$kd = 2.12$ in. and $I_{cr} = 166.8$ in.[4] (Example 5.3)

Fiber stress at the extreme tension (bottom) face when composite was applied:

$$\varepsilon_{bi} = \frac{140,000}{166.8 \times 3 \times 10^6} (12 - 2.12) = 0.0028$$

Using the results of Example 5.12, crushing of concrete occurs when extreme tension face strain is 0.0140. When this strain is reached, actual strain in the composite is:

$$\varepsilon_f = 0.0140 - \varepsilon_{bi} = 0.0140 - 0.0028 = 0.0112$$

Since this strain is less than 0.014, use this strain for computation of $a$ and $M_n$.

$$0.85 \times 5500 \times 8 \times a = 0.33 \times 60,000 + 0.85 \times 0.0112 \times 33 \times 10^6$$

$$a = 1.38 \text{ in.}$$

$$c = 1.78 \text{ in.}$$

$$M_n = 0.85 \times 5500 \times 8 \times 1.38 \left(1.78 - \frac{1.38}{2}\right) + 0.33 \times 60,000$$

$$(10 - 1.78) + 0.85 \times 0.085 \times 0.0113 \times 33 \times 10^6 (12 - 1.78)$$

$$= 494,361 \text{ in.-lb}$$

## 5.21  Ductility considerations

When beams are strengthened with composites, the failure is not initiated by yielding of tension steel. Yielding of tension steel occurs at load levels higher than normal working loads, but cannot be detected by excessive deformations as in the case of normal reinforced concrete beams. The stiffness of the beam reduces after yielding of steel but it continues to resist higher loads, producing stable increase in deflections. At the section level, the slope of the moment-curvature relationship reduces but does not approach zero. When failure occurs, either due to crushing of concrete or fracture of composite, it is more brittle than conventional reinforced concrete beam failure.

One way to ascertain a certain amount of ductility is to specify a minimum value for the ratio of curvature at failure, $\phi_u$, to curvature at yielding of steel, $\phi_y$. This ratio should increase with the increase in the ratio of nominal moment, $M_n$, to moment at yielding of steel, $M_y$. It is recommended that:

$$\text{if } \frac{M_n}{M_y} < 1.3, \text{then } \frac{\phi_u}{\phi_y} \geq 2 \qquad (5.77)$$

and if $\dfrac{M_n}{M_y} \geq 1.3$, then $\dfrac{\phi_u}{\phi_y} \geq 2.5$ (5.78)

Another option is to reduce the strength reduction factor, $\phi$, from 0.9 (recommended for flexural members, ACI code) to 0.7. The variation of $\phi$ from 0.9 to 0.7 is determined by the strain in tension steel at failure. If the strain in tension steel at failure is greater than twice the yield strain, $2\varepsilon_y$, $\phi$ is taken as 0.9. Since the steel is not allowed to yield at maximum working loads, it is assumed that loads corresponding to $2\varepsilon_y$ will result in considerable deflection and cracking. These unusual deflections and crack widths are expected to provide the warning of the impending failure.

Therefore, if $\varepsilon_s$ at failure is $\geq 2\varepsilon_y$, then $\phi = 0.9$. For other conditions:

$$\phi = 0.5 + 0.2\frac{\varepsilon_s}{\varepsilon_y} \geq 0.7 \tag{5.79}$$

## 5.22  Check for ductility requirements

*Example 5.15: Ductility of strengthened rectangular beam*
Check the ductility requirement for Example 5.14.

Solution:

The composite was installed when the maximum moment was 140,000 in.-lb. For this load:

$$f_s = \frac{140,000}{166.8} \times (10 - 2.12) \times 6.9$$

[$I_{cr}$ taken from Example 5.7]

$$= 45,600 \, \text{psi}$$

An additional stress of 14,400 psi will cause yielding of steel. For strengthened beam:

$kd = 2.5 \, \text{in.}$

$I_{cr} = 230 \, \text{in.}^4$

When $f_s = f_y$,

yield moment, $M_y = 140,000 + \dfrac{230}{(10 - 2.5) \times 6.9} \times 14,400$

$$= 140,000 + 64,000 = 204,000 \, \text{in.-lb}$$

$M_n = 489,052 \text{ in.-lb}$

$$\frac{M_n}{M_y} > 1.3$$

Curvature at yield, $\phi_y = \dfrac{\text{(yield strain of steel)}}{d - kd}$

$$= \frac{60{,}000/29 \times 10^6}{10 - 2.5} = 0.0028 \text{ rad.}$$

Curvature at ultimate, $\phi_u = \dfrac{\text{(max concrete strain)}}{\text{(depth of neutral axis)}}$

Using the results of Example 5.14:

$a = 1.38 \text{ in.}$

$c = \dfrac{1.38}{0.775} = 1.78 \text{ in.}$

$\phi_u = \dfrac{0.003}{1.78} = 0.00168 \text{ rad}$

$\dfrac{\phi_u}{\phi_y} = \dfrac{0.00168}{0.00028} = 6.0 > 2.5$

Strain in steel at failure, $\dfrac{0.003}{1.78}(10 - 1.78) = 0.014 > 2f_y \cong 0.004$

Therefore, reduction in the value of $\phi$ is not needed.

## 5.23 Sustained and fatigue loading

The effects of sustained and fatigue loading are taken into consideration when setting the limits for allowable stresses. It is assumed that concrete can withstand sustained loading. It is expected to creep, resulting in more deformations, but strength is not a concern. Steel is stable under sustained loading and essentially does not creep.

Both concrete and steel are also assumed safe under fatigue loading because the stress range (difference between maximum and minimum stresses) will be a small fraction of the total load. The endurance limit (the stress range at which the material can withstand infinite number of cycles) for steel is about 20,000 psi. Concrete can withstand about 10 million cycles at a stress range of $0.5f_c'$ (ACI Committee 215, 2000).

The response of fiber composites depends on the type of fibers. Carbon fibers can withstand a much higher percentage of their monotonic strength under sustained and fatigue loading as compared to glass fibers. It is reported that glass, aramid, and carbon fibers can sustain 0.30, 0.47, and 0.91 times their ultimate strengths, respectively (Yamaguchi *et al.*, 1997).

Using a safety factor of $1/0.6$, the following limits are recommended for the composites (ACI Committee 440, 2002). The stress due to the sustained part of the loading, $f_{fs}$, or the stress range, $f_{fr}$, must not exceed $0.20f_{fu}$, $0.30f_{fu}$, and $0.55f_{fu}$ for glass, aramid, and carbon fiber composites, respectively. The stress due to sustained load can be computed using the equation:

$$f_{fs} = \frac{(M \text{ due to sustained load})(h - kd)n_f}{I_{cr}} - E_f E_{bi} \qquad (5.80)$$

Typically, sustained load will consist of all the dead load and part of the live load. The stress range:

$$f_{fr} = \frac{(M_{max} - M_{min})(h - kd)n_f}{I_{cr}} \qquad (5.81)$$

where $M_{max}$ is the moment due to the maximum load, and $M_{min}$ is the moment due to the minimum load. Since the allowable working stress level is $0.33f_{fu}$, a separate check is not needed for carbon composites.

## 5.24 Design of flexural strengthening systems

The major steps for designing the strengthening are as follows:

*Step 1*: Obtain the details of the current slab or beam. This includes the compressive strength of concrete, location and yield strength of steel, and shear reinforcement.

*Step 2*: Estimate the loading that will be present during the application of the strengthening system. Every effort should be made to minimize the load that will be present during the application of strengthening.

*Step 3*: Choose the strengthening system. This could be carbon plates, carbon fiber sheets applied in the field, or glass fiber sheets applied in the field. Both carbon and glass fiber sheets and carbon plates are available at various strength levels.

*Step 4*: Obtain the properties of the fiber composite. This information can be obtained from the manufacturers.

*Step 5*: Estimate the moment capacity of the existing beam (slab).

*Step 6*: Estimate the moment capacity needed for upgraded loads.

*Step 7*: Estimate the amount of composite reinforcement needed. The strength computation equations will provide an approximate area. This area should be converted into plate thickness and width or number of layers and width of sheets (fabrics). This step can be considered as preliminary design.

*Step 8*: The composite should generate sufficient additional moment capacity to resist the extra loads. In some instances, composite reinforcement will be provided to correct the deficiencies. For example, reinforcement lost to corrosion could be replaced with composite reinforcement.

*Step 9*: Check to ascertain that the working load moment for the upgraded loads does not exceed the following stresses:

Maximum concrete stress, $0.45f_c'$
Maximum steel stress in tension, $0.8f_y$
Maximum steel stress in compression, $0.4f_y$
Maximum composite stress, $0.33f_u$

Depending on the type and exposure, the allowable composite stresses might have to be reduced further.

*Step 10*: Check for ductility requirements.

*Step 11*: Check for creep rupture and fatigue stress limits.

*Step 12*: Check for deflection limits.

### 5.24.1   *Preliminary design (estimation of composite area)*

An initial estimate of the composite area can be made using the following guidelines. Once the area of composite is established, detailed analysis should be carried out to check all the requirements at working and ultimate load.

*Step 1*: Compute the design ultimate moment, $M_u$, based on the upgraded loads.

*Step 2*: Compute $\phi M_{ni}$ of unstrengthened cross-section.

*Step 3*: Assuming a failure strain of about $0.8\varepsilon_{fu}$ for composite, and a lever arm of $0.9h$, estimate the area of composite, $A_f$:

$$A_f = \frac{M_u - \phi M_{ni}}{\phi(E_f 0.8\varepsilon_{fu} \times 0.9h)} \tag{5.82}$$

*Step 4*: Decide the width of the composite. Normally, the width of the beam will control the width of the composite. In the case of slabs, 3- or 4-in. wide strip could be used for each 12-in. width of slab.

*Step 5*: Estimate the thickness of the plate or the number of layers of sheet reinforcement. The width and number of layers can be adjusted simultaneously to obtain round members. In the case of multiple layer application, it is advisable to use the same width for all the layers.

### 5.24.2   *Final design*

Once the area of composite is determined, the properties of the entire section are known for checking the various requirements. The following sequence of calculations can be used as a guideline.

*Step 1*: For the unstrengthened section, obtain the properties of uncracked section and cracking moment. These include the computations of $E_c, \overline{y}, I_g$, and $M_{cr}$.

*Step* 2: Compute the properties of cracked, unstrengthened section. These include $kd$ and $I_{cr}$.

*Step* 3: For the loads present at the installation of composite, compute $f_c$, $f_s$, $\varepsilon_{bi}$ (extreme tension fiber), and $\delta$. Note that, for this step, the properties of the unstrengthened section should be used.

*Step* 4: Compute $\phi M_n$ for the strengthened section.

*Step* 5: If $\phi M_n \geq M_u$, proceed further.

*Step* 6: Otherwise, revise the area of composite and recompute $\phi M_n$.

*Step* 7: Check for ductility requirements.

*Step* 8: Compute $f_c$, $f_s$, and $f_f$ for the revised loads. Note that the stresses for the difference between the original and upgraded loads (moments) should be computed using the properties of the strengthened section and added to the stresses obtained in Step 3.

*Step* 9: If allowable stresses for worked load are satisfied, proceed to *Step 8*. Otherwise, revise the area of composite and repeat Steps 4–6, or just *Step 6*.

*Step* 10: Compute the deflection and check for allowable limits.

*Step* 11: Check for other strength parameters such as shear.

## 5.25 Design examples: flexural strengthening

*Example 5.16: Bridge slab strengthened with carbon composite*

Design a strengthening system for the following 70-year-old, solid slab, concrete bridge. Estimated concrete strength is 3000 psi, yield strength of steel is 30,000 psi. The effective depth and total thickness are 16.5 and 18.5 in., respectively. The slab bridge has to be strengthened to carry a working (service) load bending moment of 504,000 in.-lb per 12 in. of width and a factored load design moment, $M_u$, of 792,000 in.-lb per 12 in. of width. The concrete is in good condition and there are no signs of corrosion of reinforcement. During the installation of composite, the live loads will be removed. Estimated working load moment, $M_i$, is 240,000 in.-lb per 12 in. of width. Area of steel per 12-in. width of slab is 1.5 in.$^2$. The simply supported span is 19 feet.

Solution:

A 12-in. wide strip is considered for design.

Preliminary design:

Compute the nominal capacity of unstrengthened section:

$$a = \frac{A_s f_y}{0.85 f_c' b} = \frac{1.5 \times 30,000}{0.85 \times 3000 \times 12} = 1.47 \, \text{in.}$$

$$\phi M_{ni} = 0.9 \times 1.5 \times 30,000 \, (16.5 - 1.47/2)$$

$$= 638,500 \, \text{in.-lb}$$

Try carbon composite with a fiber modulus of $33 \times 10^6$ psi and fracture strength of 550,000 psi.
Therefore:

$$\varepsilon_{fu} = 0.017$$

Area of composite needed, $A_f = \dfrac{M_u - \phi M_{ni}}{\phi(E_f 0.8\varepsilon_{fu}^* \times 0.9b)}$

$$= \frac{792,000 - 638,500}{0.9(33 \times 10^6 \times 0.8 \times 0.017 \times 0.9 \times 18.5)}$$

$$= 0.023 \text{ in.}^2$$

Equivalent thickness of one layer is 0.0065 in.

Therefore, width of sheet needed per 12-in. width of slab is, $\dfrac{0.023}{0.0065} =$ 3.54 in.

Try one layer (ply), 4-in wide for every 12 in. of slab width or 4-in.-wide strip distributed at 12 in. center-to-center for the entire width of the bridge.
Therefore area of composite fiber, $A_f = 4 \times 0.0065 = 0.026$ in.
Properties of uncracked, unstrengthened section:

Depth of neutral axis, $\overline{y} = \dfrac{h}{2} = \dfrac{18.5}{2} = 9.25$ in.

Gross moment of inertia, $I_g = \dfrac{bh^3}{12} = \dfrac{12 \times 18.5^3}{12} = 6332$ in.$^4$

Modulus of rupture, $f_r = 7.5\sqrt{f_c'} = 7.5\sqrt{3000} = 411$ psi

Cracking moment, $M_{cr} = \dfrac{I_g}{h - \overline{y}} f_r = \dfrac{6332}{18.5 - 9.25} \times 411 = 281,329$ in.-lb

Experimental cracking moment is about 250,000 in.-lb.

Properties of cracked, unstrengthened section:

$$E_c = 57,000\sqrt{3000} = 3.12 \times 10^6 \text{ psi}$$

Assuming $E_s = 29 \times 10^6$ psi, $n = 29 \times 10^6 / 3.12 \times 10^6 = 9.3$

$$\frac{b(kd)^2}{2} = nA_s (d - kd)$$

$$\frac{12(kd)^2}{2} = 9.3 \times 1.5 (16.5 - kd)$$

$$kd = 5.14 \text{ in.}$$

$$I_{cr} = \frac{b(k_d)^3}{3} + nA_s(d - k_d)^2$$

$$= \frac{12(5.14)^3}{3} + 9.3 \times 1.5 \,(16.5 - 5.14)^2 = 2343 \text{ in.}^4$$

For the moment that is present at installation, 240,000 in.-lb:

$$f_c = \frac{240,000}{2343} \times 5.14 = 527 \text{ psi}$$

$$\varepsilon_c = \frac{527}{3.12 \times 10^6} = 0.00017$$

$$f_s = \frac{240,000}{2343} \,(16.5 - 5.14) \times 9.3 = 10,822 \text{ psi}$$

$$\varepsilon_s = 0.00037$$

Strain at the extreme bottom fiber using similar triangles:

$$\varepsilon_{bi} = \frac{18.5 - 5.14}{5.14} \times 0.00017 = 0.00044 \text{ in./in.}$$

Nominal moment capacity of strengthened section:
  Assuming failure is initiated by concrete:

$$0.85 f_c' b \beta_1 c = A_s f_y + A_f E_f 0.003 \left(\frac{h - c}{c}\right)$$

$$0.85 \times 3000 \times 12 \times 0.85 \times c = 1.5 \times 30,000 + 0.026 \times 33 \times 10^6$$

$$\times 0.003 \left(\frac{18.5 - c}{c}\right)$$

$$c^2 - 1.6c - 18.3 = 0$$

$$c = 2.4 \text{ in.}$$

$$\varepsilon_f = 0.003 \left(\frac{18.5 - 2.4}{2.4}\right) = 0.02 \text{ in./in.}$$

For fracture of composite, strain at bottom:

$$\varepsilon_{bi} + \varepsilon_{fu} = 0.017 < 0.002$$

Therefore, failure is initiated by fracture of composite. However, the strain values are close and, hence, rectangular stress distribution can be assumed for concrete.

$$nt_f E_f = 0.0065 \times 1 \times 33 \times 10^6 = 214{,}500$$

$$\kappa_m = 1 - \frac{214{,}500}{2{,}400{,}000} = 0.91$$

$$\kappa_m = 0.9$$

$$\varepsilon_{fu} = 0.015$$

$$0.85 \times 3000 \times 12 \times a = 1.5 \times 30{,}000 + 0.026 \times 33 \times 10^6 \times 0.0153$$

$$a = 1.90 \text{ in.}$$

$$c = 1.90/0.85 = 2.3 \text{ in.}$$

$$M_n = 1.5 \times 30{,}000\,(16.5 - 1.90/2) + 0.85 \times 0.026 \times 0.0153$$

$$\times 33 \times 10^6 \times (18.5 - 1.90/2)$$

$$= 895{,}578 > \frac{M_u}{\phi} = \frac{792{,}000}{0.9} = 880{,}000 \text{ in.-lb}$$

A nonlinear analysis results in $a = 1.94$ in., $\beta = 0.833$, and $M_n = 948{,}500$ in.-lb with an error of 2%.

Long-term effects:
For long-term exposure, the allowable maximum stress (strain) has to be reduced further. For structures exposed to outside environment, the reduction factor is 0.85.

Therefore, $\varepsilon_{fu} = 0.85 \times 0.0153 = 0.0130$

The failure is still by fracture of composite.

$$0.85 \times 3000 \times 12 \times a = 1.5 \times 30{,}000 + 0.026 \times 33 \times 10^6 \times 0.0138$$

$$a = 1.86 \cong 1.90 \text{ in.}$$

Therefore, $M_n \cong 895{,}578$ in.-lb. Hence OK.

The difference is not significant because the contribution of the composite is small as compared to the contribution of steel.
Check for ductility:

Strain in carbon, $\varepsilon_f = 0.0153$

Strain in tension steel, $= (\varepsilon_f + \varepsilon_{bi}) \left( \dfrac{d - c}{h - c} \right)$

$$= (0.0153 + 0.00044) \left( \frac{16.5 - 2.3}{18.5 - 2.3} \right)$$

$$= 0.0138 > 2\varepsilon_y = 0.0041$$

Therefore $\phi = 0.9$

Hence, the chosen area of composite satisfies strength requirements.
Check for allowable stresses at working loads:

$b = 12$ in.

$d = 16.5$ in.

$h = 18.5$ in.

$A_s = 1.5$ in.$^2$

$A_f = 0.026$ in.$^2$

$n_f = E_f/E_c = 33 \times 10^6/3.12 \times 10^6 = 10.6$

For cracked section:

$$\frac{12(kd)^2}{2} = 9.3 \times 1.5\,(16.5 - kd) + 0.026 \times 10.6\,(18.5 - kd)$$

$$(kd)^2 + 2.37kd - 39.21 = 0$$

$$kd = 5.19 \text{ in.}$$

$$I_{cr} = \frac{12(5.19)^3}{3} + 9.3 \times 1.5\,(16.5 - 5.19)^2 + 10.6 \times 0.026$$

$$(18.5 - 5.19)^2 = 2391 \text{ in.}^4$$

Working load bending moment is 504,000 in.-lb.

$$f_c = \frac{504,000}{2392} \times 5.12 = 1078 \text{ psi}$$

$$= 0.36f_c' < 0.45f_c'$$

$$\varepsilon_c = \frac{1078}{3.12 \times 10^6} = 0.00346$$

$$f_s = \frac{504,000}{2392}\,(16.5 - 5.12) \times 9.3 = 22,300 \text{ psi}$$

$$= 0.74f_y < 0.8f_y$$

$$f_f = 33 \times 10^6 \times \text{strain at bottom face}$$

$$= 33 \times 10^6 \left[ \varepsilon_c \left( \frac{h - kd_d}{kd} \right) - \varepsilon_{bi} \right]$$

$$= 33 \times 10^6 \left( 0.000346 \times \frac{18.5 - 5.12}{5.12} - 0.00044 \right) = 15,318 \text{ psi}$$

$$= 0.028f_{fu} < 0.33f_{pu}. \text{ Therefore OK.}$$

Check for deflection:

$$I_c = 2392 + (6332 - 2392)\left(\frac{281,329}{504,000}\right)^3 = 3077\,\text{in.}^4$$

Assuming that the moment is produced by uniformly distributed load,

$$\delta = \frac{5ML^2}{48E_cI_e} = \frac{5 \times 504,000 \times (19 \times 12)^2}{48 \times 3.12 \times 10^6 \times 3077} = 0.28\,\text{in.}$$

$$\delta = \frac{L}{814} < \frac{L}{800}$$

### Example 5.17: Rectangular beam with compression and tension reinforcement

Design the strengthening system using the following information: width of beam, $b = 12$ in.; thickness, $h = 20$ in.; effective depth of tension steel, $d = 16.5$ in.; effective depth for compression steel, $d' = 2.5$ in.; area of tension steel, $A_s = 3.6$ in.$^2$ (6 – #7 bars); area of compression steel, $A_s' = 0.88$ in.$^2$ (2 – #6 bars). Simply supported beam over a span of 24 feet.

Current loads:

$w_{LL} = 650\,\text{lb/ft}$

$w_{DL} = 1450\,\text{lb/ft}$

$w_{WL} = 2100\,\text{lb/ft}$

$w_{UL} = 3135\,\text{lb/ft}$

Projected (upgraded) loads:

$w_{LL} = 1100\,\text{lb/ft}$

$w_{DL} = 1450\,\text{lb/ft}$

$w_{WL} = 2550\,\text{lb/ft}$

$w_{UL} = 3900\,\text{lb/ft}$

Compressive strength of concrete, $f_c' = 4000$ psi
Yield strength of steel, 60,000 psi
Modulus of elasticity of steel, $29 \times 10^6$ psi
Modulus of elasticity of carbon fibers, $33 \times 10^6$ psi
Fracture strain, $\varepsilon_{fu} = 0.017$
Equivalent fiber thickness, 0.0065 in.

Assume that only dead load is present during renovation.

Solution:

Preliminary design:

Compute the nominal moment capacity of the unstrengthened section. Neglecting the contribution of compression steel:

$$a = \frac{A_s f_y}{0.85 f_c' b} = \frac{3.6 \times 60,000}{0.85 \times 4000 \times 12} = 5.30 \, \text{in.}$$

For $f_c' = 4000 \, \text{psi}$, $\beta_1 = 0.85$

$$c = \frac{a}{\beta_1} = \frac{5.30}{6.23} = 6.23 \, \text{in.}$$

Strain at compression level, $\varepsilon_s' = 0.003 \left( \frac{c - d'}{d} \right) = 0.003 \left( \frac{6.23 - 2.5}{6.23} \right)$

$$= 0.0018$$

$$f_s' = 0.0018 \times 29 \times 10^6 = 52,066 \, \text{psi}$$

With inclusion of compression steel force, the depth of neutral axis and $f_s'$ will decrease. Therefore, assume $f_s' = 45,000 \, \text{psi}$.

For next iteration, $a = \dfrac{3.6 \times 60,000 - 0.88 \times 45,000}{0.85 \times 4000 \times 12} = 4.32 \, \text{in.}$

$c = 5.09 \, \text{in.}$

$$f_s' = 29 \times 10^6 \times 0.003 \left( \frac{5.09 - 2.5}{5.09} \right) = 44,209 \, \text{psi}$$

Therefore, the stress in compression steel is between 44,209 and 45,000 psi.

Assume $f_s' = 44,600 \, \text{psi}$.

$$a = \frac{3.6 \times 60,000 - 0.88 \times 44,600}{0.85 \times 4000 \times 12} = 4.33 \, \text{in.}$$

Check for minimum reinforcement ratio:

$$\frac{200}{f_y} = 0.0033$$

$$\frac{3\sqrt{f_c'}}{f_y} = \frac{3\sqrt{4000}}{60,000} = 0.00316$$

$$\frac{A_s}{bd} = \frac{2.5}{12 \times 16.5} = 0.018 > 0.00333 > 0.00316$$

Check for maximum reinforcement ratio:

$$\beta_1 = 0.85$$

$$0.75\beta_1 \left(\frac{0.85f_c'}{f_y}\right)\frac{87,000}{87,000+f_y}$$

$$= 0.75 \times 0.85 \left(\frac{0.85 \times 4000}{60,000}\right)\frac{87,000}{87,000+60,000}$$

$$= 0.0213$$

$$\frac{A_s}{bd} - \frac{A_s'}{bd}\left(\frac{f_s'}{f_y}\right) = \frac{3.6}{12 \times 16.5} - \frac{0.88}{12 \times 16.5}\left(\frac{44,600}{60,000}\right)$$

$$= 0.0151 < 0.0213$$

$$\phi M_{ni} = 0.9\left[0.85f_c'ba\left(d - \frac{a}{2}\right) + A_s'f_s'\left(d - d'\right)\right]$$

$$= 0.9\left[\begin{array}{l} 0.85 \times 4000 \times 12 \times 4.33\left(16.5 - \dfrac{4.33}{2}\right) \\ +0.88 \times 44,600\,(16.5 - 2.5) \end{array}\right]$$

$$= 2{,}773{,}755\ \text{in.-lb}$$

Area of composite needed, $A_f = \dfrac{M_u - \phi M_{ni}}{\phi(E_f 0.8\varepsilon_{fu}^* \times 0.9h)}$

For the revised loads, $M_u = \dfrac{w_{UL}L^2}{8} = \dfrac{3900 \times 24^2 \times 12}{8}$

$$A_f = \frac{3{,}369{,}600 - 2{,}773{,}755}{0.9(33 \times 10^6 \times 0.8 \times 0.017 \times 0.9 \times 20)}$$

$$= 0.084\ \text{in.}^2$$

For a fiber sheet thickness of 0.0065 in.

Width required, $\dfrac{0.084}{0.0065} = 12.9\ \text{in.}$

Since the sheet comes in 20-in. width, try three layers of 10-in. width sheets.

Therefore, $A_f = 3 \times 10 \times 0.0065 = 0.195\ \text{in.}^2$

Extra area is needed to compensate for loss of fracture strain due to multiple layers.

Properties of uncracked section:

$$I_g = \frac{bh^3}{12} = \frac{12 \times 20^3}{12} = 8000\ \text{in.}^4$$

$$M_{cr} = \frac{I_g}{h/2} f_r = \frac{8000}{6} \times 7.5\sqrt{4000}$$

$$= 632{,}456 \text{ in.-lb}$$

$$E_c = 57{,}000\sqrt{f_c'} = 57{,}000\sqrt{4000} = 3.6 \times 10^6 \text{ psi}$$

Properties of cracked unstrengthened section:

$$\frac{b(kd)^2}{2} + (n-1)A_s'(kd-d) = nA_s(d-kd)$$

$$n = 29 \times 10^6 / 57{,}000\sqrt{f_c'} = 8.1$$

$$6(kd)^2 + 7.1 \times 0.88(kd-2.5) = 8.1 \times 3.6(16.5-kd)$$

$$6(kd)^2 + 35.4kd - 496.8 = 0$$

$$kd = 6.6 \text{ in.}$$

$$I_{cr} = \frac{bc^3}{3} + nA_s(d-k_d)^2 + (n-1)A_s'(kd-d')^2$$

$$\frac{12 \times 6.6^3}{3} + 8.1 \times 3.6(16.5-6.6)^2 + 7.1 \times 0.88(6.6-2.5)^2$$

$$= 4113 \text{ in.}^4$$

Moment at installation, $\dfrac{1450 \times 24^2}{8} \times 12 = 1{,}461{,}600 \text{ in.-lb}$

$$\varepsilon_{ci} = \frac{1{,}461{,}600}{4113} \times \frac{6.6}{3.6 \times 10^6} = 0.00065$$

$$\varepsilon_{bi} = 0.00065\left(\frac{20-6.6}{6.6}\right) = 0.00132$$

Nominal moment capacity of the strengthened beam:
Assume failure is initiated by the crushing of concrete. Assume $f_s' = 50{,}000$ psi. Note that neutral axis will move down, increasing strain in compression steel.

$$0.85f_c'b\beta_1 c + A_s'f_s' = A_sf_y + A_fE_f0.003\left(\frac{h-c}{c}\right)$$

$$0.85 \times 4000 \times 12 \times 0.85 \times c = 3.6 \times 60{,}000 + 0.195 \times 33 \times 10^6$$

$$\times 0.003\left(\frac{20-c}{c}\right)$$

$$34.7c^2 + 44c = 216c - 19.3c + 386.1$$

or $c^2 - 4.4c - 11.12 = 0$

$c = 6.20\,\text{in.}$

$a = 5.27\,\text{in.}$

$$\varepsilon_s' = 0.003 \left( \frac{6.2 - 2.5}{6.2} \right) = 0.00179$$

$$f_s' = 0.00179 \times 29 \times 10^6 = 51{,}919\,\text{psi}$$

Use a revised $f_s'$ for the next iteration, Figure 5.22:

$$\varepsilon_f = 0.003 \left( \frac{20 - 6.2}{6.2} \right) = 0.0067$$

$$\varepsilon_{bi} + \varepsilon_{fu} = 0.017 + 0.00132 = 0.018$$

$$\varepsilon_f = \varepsilon_{bi} + \varepsilon_{fu}$$

Therefore, failure will occur due to crushing of concrete.

$$nt_f E_f = 3 \times 0.0065 \times 33 \times 10^6 = 643{,}500$$

$$\kappa_m = 1 - \frac{643{,}500}{2{,}400{,}000} = 0.732$$

$$\varepsilon_{fu} = 0.732 \times 0.017 = 0.0124$$

If we use this strain instead of $\varepsilon_{fu}$:

$$\varepsilon_{bi} + \varepsilon_{fu} = 0.00132 + 0.0124 = 0.0137$$

This strain is greater than 0.0067 required to cause crushing of concrete.

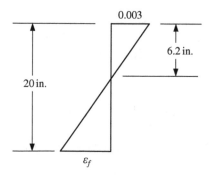

*Figure 5.22* Strain distribution for beam in Example 5.17.

Therefore, failure occurs by crushing of concrete. If the stress in compression steel is increased to 51,500 psi, the neutral axis will slightly decrease.

If we assume, $c = 6.15$ in.

$a = 5.23$ in.

Compressive force, $0.85 f_c' ba + A_s' f_s'$

$= 0.85 \times 4000 \times 12 \times 5.23 + 0.88 \times 51,500$

$= 258,704$ lb

Tension force, $A_s f_y + A_f E_f 0.003 \left( \dfrac{20 - 6.15}{6.15} \right)$

$= 3.6 \times 60,000 + 0.195 \times 33 \times 10^6 \times 0.003 \left( \dfrac{13.85}{6.15} \right)$

$= 259,476$ lb

$C \cong T$

Therefore, $a = 5.23$ in.

$c = 6.15$ in.

$f_s' = 51,000$ psi

$E_f \varepsilon_f = f_f = 222,951$ psi

$M_n = 0.85 f_c' ba (c - a/2) + A_s' f_s' (c - d') + A_s f_y (d - c)$

$\quad\quad + 0.85 A_f f_f (h - c)$

$= 0.85 \times 4000 \times 12 \times 5.23 (6.15 - 5.23/2) + 0.88$

$\quad \times 51,500 (6.15 - 2.5) + 3.6 \times 60,000 (16.5 - 6.15) + 0.85$

$\quad \times 0.195 \times 222,951 (20 - 6.15)$

$= 3,667,145$ in.-lb

$\phi M_n = 3,300,430 < M_u = 3,369,600$ in.-lb

Even though $\phi M_n$ is slightly less than $M_u$, further checks are being carried out for long-term exposure if fracture strain is reduced by 0.9.

$\varepsilon_{fu} = 0.9 \times 0.0124 = 0.01116$

This strain is still larger than strain at failure, 0.0067. Therefore, no new check is needed.

Check for allowable stresses at working load:

$b = 12\,\text{in.}$

$d = 16.5\,\text{in.}$

$h = 20\,\text{in.}$

$A_s = 3.6\,\text{in.}^2$

$A_s' = 0.88\,\text{in.}^2$

$A_f = 0.195\,\text{in.}^2$

$E_c = 3.6 \times 10^6\,\text{psi}$

$n = 8.1$

$n_f = E_f / E_c = 33 \times 10^6 / 3.6 \times 10^6 = 9.2$

For cracked section:

$$6\,(kd^2) + 7.1 \times 0.88\,(kd - 2.5) = 8.1 \times 3.6\,(16.5 - kd) + 0.195$$
$$\times\, 9.2\,(20 - kd)$$

$$6(kd)^2 + 6.25kd - 15.62 + 29.16kd - 481.1 + 1.8kd - 35.9 = 0$$

$$\text{or,}\ (kd)^2 + 6.2kd - 88.8 = 0$$

$$kd = 6.82\,\text{in.}$$

$$I_{cr} = \frac{12(6.82)^3}{3} + 7.1 \times 0.88\,(6.82 - 2.5)^2$$
$$+\, 8.1 \times 3.6\,(16.5 - 6.82)^2 + 0.195 \times 9.2\,(20 - 6.82)^2$$
$$= 4429\,\text{in.}^4$$

Working load bending moment, $\dfrac{2550 \times 24^2}{8} \times 12 = 2{,}203{,}200\,\text{in-lb}$

$$f_c = \frac{2{,}203{,}200}{4429} \times 6.82 = 3393\,\text{psi}$$
$$= 0.848f_c' > 0.45f_c'$$

Since the stress level is too high, a large number of composite layers will be required. Therefore, it is recommended not to use composites for upgrades. Note that the original section satisfies strength requirements, but not stresses at working loads. For the unstrengthened section, $f_c > 0.45f_c'$ for the working load.

For unstrengthened section:

$\phi M_n = 2,773,755 \text{ in.-lb}$

$M_u = 2,708,640 < \phi M_n$

For total working load of 2100 lb/ft,

$M = 1,814,400 \text{ in.-lb}$

$f_c = 2912 = 0.727f_c' > 0.45f_c'$

## 5.26 Problems

**Problem 5.1:** Compute the nominal moment capacity of a rectangular beam strengthened with two plies that are 12-in. wide. Beam width is 12 in., depth is 21.5 in., thickness is 24 in. Area of steel (3 – ASTM No. 9 bars) = 3 in². Compressive strength of concrete = 5000 psi, yield strength of steel = 60,000 psi, modulus of steel = $29 \times 10^6$ psi. Modulus of composite (fiber only) = $33 \times 10^6$ psi, thickness of composite (fiber only) = 0.0065 in. Guaranteed rupture strain for composite = 0.017. The composite is attached to the bottom of the beam.

**Problem 5.2:** The beam of Example 5.1 is used to support a working dead and live loads of 1000 and 1800 lb/ft, respectively. The beam is simply supported over a span of 24 ft. The strengthening was applied when a load of 1000 lb/ft was present. Check whether the section satisfies ultimate and working level conditions. Compute the maximum deflection.

**Problem 5.3:** Estimate the carbon fiber area required for a slab to withstand the revised loads. Simply supported span = 9 ft, thickness of the slab = 6 in., effective depth = 4 in., uniformly distributed dead load = 100 lb/ft, uniformly distributed live load = 1100 lb/ft. Compressive strength of concrete = 3000 psi, yield strength of steel = 40,000 psi, modulus of elasticity of steel = $29 \times 10^6$ psi, fracture strain of carbon fibers = 0.0167. The sheets are available in 20-in. width with equivalent fiber thickness 0.0065 in. Reinforcement: ASTM, No. 4 bars at a spacing of 8 in. or area of steel for 12-in. width of slab = 0.3 in.². Load at the time of application of load = 200 lb/ft. Exposure condition: inside closed area.

**Problem 5.4:** Compute the carbon area needed for the following rectangular beam. Width = 12 in., depth = 32 in., thickness = 36 in., area of steel: 2.65 in.² (6 – #6 bars). Simply supported span of 48 feet, existing dead load = 550 lb/ft, and existing live load = 350 lb/ft. Upgraded dead load = 560 lb/ft, upgraded live load = 500 lb/ft. Assume a dead load of 550 lb/ft when carbon sheets are applied. Compressive strength of concrete = 4000 psi, yield strength of steel = 60,000 psi, fracture strain of carbon = 0.0167, sheet

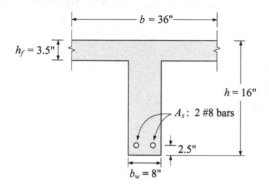

*Figure 5.23* T-beam associated with Problem 5.7.

thickness (fibers only) = 0.0065 in., fiber modulus = $33 \times 10^6$ psi. Exposure: outside environment. $f_c' = 4000$ psi, $f_y = 60,000$ psi, $E_s = 29 \times 10^6$ psi, $E_f = 33 \times 10^6$ psi, $\varepsilon_{fu} = 0.0167$. For the current loads, $M_u = 1,200,000$ in.-lb. For the upgraded loads, $M_u = 140,000$ in.-lb.

**Problem 5.5:** Repeat problem 5.2 using high-modulus carbon fiber with a modulus of $54 \times 10^6$ psi and fracture strain of 0.008.

**Problem 5.6:** Due to expansion of a pumping facility, the beams need strengthening. Investigate whether carbon sheets could be used for increasing load capacity. Fiber fracture strain = 0.0167, fiber modulus = $33 \times 10^6$ psi, beam width = 16 in., thickness = 30 in., and depth = 27 in. Tension steel area = 3.95 in.$^2$ (5 – #7 bars). Compression steel area = 0.88 in.$^2$ (2 – #6 bars). Current dead load = 1600 lb/ft and live load = 1700 lb/ft. Revised dead load = 1600 lb/ft and live load = 2250 lb/ft. Assume only dead load is present during the application of carbon. Compressive strength of concrete = 6000 psi. Simply supported span = 26 ft.

**Problem 5.7:** Obtain the carbon area needed for the T-beam shown in Figure 5.23.

# References

ACI Committee 215 (2000) *Considerations for Design of Concrete Structures Subjected to Fatigue Loading (ACI 215R-74) (Revised 1997)*. Farmington Hills, MI: American Concrete Institute, 24 pp.

ACI Committee 318 (2005) *Building Code Requirements for Structural Concrete and Commentary (ACI 318R-05)*. Farmington Hills, MI: American Concrete Institute, 430 pp.

ACI Committee 440 (1996) *State-of-the-Art Report on Fiber Reinforced Plastic (FRP) Reinforcement for Concrete Structures (ACI 440 R-96)*. Farmington Hills, MI: American Concrete Institute, 68 pp.

ACI Committee 440 (2002) *Guide for the Design and Construction of Externally Bonded FRP Systems for Strengthening Concrete Structures (ACI 440.2R-02)*. Detroit, MI: American Concrete Institute, 45 pp.

Arduini, M., Di Tommaso, A., and Nanni, A. (1997) Brittle failure in FRP plate sheet bonded beams. *ACI Structural Journal*, 94(4), pp. 363–370.

Crasto, A.S., Kim, R.Y., Fowler, C., and Mistretta, J.P. (1996) Rehabilitation of concrete bridge beam with externally bonded composite plates. In: *Proceedings of the First International Conference on Fiber Composites in Infrastructure*, Arizona, USA, January 1996. Netherlands: Springer, pp. 814–828.

De Lorenzis, L. and Nanni, A. (2001) Characterization of FRP rods as near-surface-mounted reinforcement. *Journal of Composites for Construction*, 5(2), pp. 114–121.

Kurtz, S. (2000) *Analysis and Prediction of Delamination of Strengthened Beams*. PhD thesis, New Brunswick, NJ: Rutgers, The State University of New Jersey, 610 pp.

Limbrunner, G.F. and Aghayere, A.O. (2007) *Reinforced Concrete Design*. 6th edn. Upper Saddle River, NJ: Pearson Prentice Hall, 521 pp.

MacGregor, J.G. and Wight, J.K. (2005) *Reinforced Concrete: Mechanics and Design*. 4th edn. Upper Saddle River, NJ: Pearson Prentice Hall, 1132 pp.

Malek, A., Saadathmanesh, H., and Ehsani, M. (1998) Prediction of failure load of RC beams strengthened with FRP plate due to stress concentrations at the plate end. *ACI Structural Journal*, 95(1), pp. 142–152.

M'Bazaa, I., Missihoun, M., and Labossiere, P. (1996) Strengthening of reinforced concrete beams with CFRP sheets. In: *Proceedings of the First International Conference on Fiber Composites in Infrastructure*, Arizona, USA, January 1996. The Netherlands: Springer, pp. 746–759.

Nanni, A. and Faza, S. (2002) Designing and constructing with FRP bars: an emerging technology. *Concrete International*, 24(11), pp. 53–58.

Nawy, E.G. (2009) *Reinforced Concrete: A Fundamental Approach*. 6th edn. Upper Saddle River, NJ: Pearson Prentice Hall, 915 pp.

Setareh, M. and Darvas, R. (2007). *Concrete Structures*. 1st edn. Upper Saddle River, NJ: Pearson Prentice Hall, 564 pp.

Todeschini, C., Bianchini, A., and Kesler, C. (1964) Behavior of concrete columns reinforced with high-strength steels, *ACI Journal, Proceedings*, 61(6), pp. 701–716.

Wang, C., Salmon, C.G., and Pincheira, J.A. (2007) *Reinforced Concrete Design*. 7th edn. New Jersey: John Wiley & Sons, 948 pp.

Yamaguchi, T., Kato, Y., Nishimura, T., and Uomoto, T. (1997) Creep rupture of FRP rods made of aramid, carbon and glass fibers. In: *Proceedings of the Third International Symposium on Non-metallic (FRP) Reinforcement for Concrete Structures (FRPRCS-3)*, Sapporo, Japan, Vol. 2, October 1997. Tokyo, Japan: Japan Concrete Institute, pp. 179–186.

# 6 Flexure
## Prestressed concrete

## 6.1 Introduction

In reinforced concrete, concrete resists compressive forces whereas reinforcing steel resists the tensile forces. In prestressed concrete beams, the contribution of concrete is maximized by precompressing it with high-strength steel bars, wires, or cables. The ultimate strength of prestressed cables could be as high as 270,,000 psi as compared to 90,000 psi for typical steel used for reinforcement. By optimizing the prestress and the shape, beams can be designed so that they do not crack under normal service loads. This allows for the utilization of the entire cross-section resulting in much smaller deflections. The prestress can also be designed to compensate dead load so large spans can be covered.

There are two general classes of prestressed concrete beams. If concrete is cast around pretensioned wires and the wires are cut after concrete cures, they are called pretensioned beams. If the beams are stressed after they are cured, they are called post-tensioned beams. Typically, ducts left in beams during casting are utilized for placing prestressing cables. These cables, prestressed against the beams at the ends, induce the prestress. If the ducts are grouted, then the prestressed reinforcement binds with the concrete and they are called post-tensioned beams with bonded tendons. These beams behave similarly to the pretensioned beams except for prestress losses and stresses induced in nonprestressed reinforcement. In both pretensioned beams and post-tensioned beams with bonded tendons, the properties are section dependent. The change in stress in the prestressed cable depends on the curvature of the section.

If the cables are left ungrouted, they are called prestressed beams with unbonded tendons. In this case, the change in stress in the prestressed cable is not section dependent. In other words, the change in prestress depends upon the geometry of the beam, cable geometry, span, and support condition. The deflected shape of the entire beam controls the change in stress.

Prestressed beams are commonly used for long span structures. These include bridges, industrial buildings, and parking garages. Parking garages are prime candidates for repair and rehabilitation because the deicing salts used to melt ice create a very favorable condition for corrosion. Accidents can damage webs of double-tee cross-sections commonly used for garages. Unaccounted loss of prestress, say due to insufficient strength gain when pretensioned wires are cut from bulkhead, could also result in reduction of load-carrying capacity. Any deterioration in the end (near supports) could result in slippage and loss of prestress. All these factors contribute to frequent rehabilitation needed in these structures. High-strength composites, commonly known as fiber-reinforced polymers (FRP), have been utilized successfully in a number of projects (Rosenboom *et al.*, 2004, 2008).

## 6.2  Behavior of typical prestressed concrete beam

Typical load–deflection behavior of a prestressed concrete beam is shown in Figure 6.1. The behavior has both similarities and significant differences as compared to the behavior of reinforced concrete presented earlier in Figure 5.1. The following are the similarities:

- Both load–deflection curves have three segments representing uncracked, cracked – preyielding, and cracked – post-yielding segments.
- As the cracking becomes more prevalent, stiffness of the beam decreases.

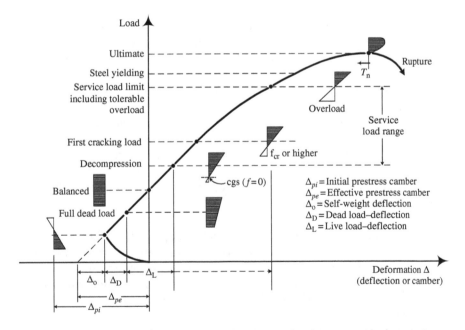

*Figure 6.1* Typical load–deflection behavior of a prestressed concrete beam (Source: Nawy, Edward G., *Prestressed Concrete: A Fundamental Approach*, 4th edn, © 2003, p. 109. Reprinted by permission of Pearson Education, Inc., Upper Saddle River, NJ).

- For typical under-reinforced sections, both beams provide considerable post-yielding deflections and, hence, ample warning of the impending failure.
- In terms of analysis, once the concrete cracks, the tension force contribution of concrete is neglected.

The major differences are in the overall stiffness and load-carrying capacity as discussed below.

- Since the concrete is prestressed, it does not crack until a certain amount of external load is applied. In most cases, the beam may not crack under service load and, hence, the entire section provides flexural stiffness reducing the deflection. In addition, the prestressing force typically induces deflection in the opposite direction of external load. This deflection, known as camber, further reduces the final deflection.
- Prestressing also increases the strength capacity of the beams because the prestressing force induces a moment that balances part of the external load.
- When tension zone concrete cracks, the loss of stiffness (or increase in deflection) is more gradual in prestressed concrete beams because the prestress provides for a more stable and controlled crack growth toward the neutral axis.
- The loss of flexural stiffness due to yielding of nonprestressed steel is also more gradual because prestressed reinforcement, which yields at much larger strains, continues to provide increase in tension force with increase in curvature.
- Post-yielding increase in both strength and deflection are larger for prestressed concrete beams.
- All the advantages provided by prestressed reinforcement become negative if there is a loss of prestress due to corrosion, loss of bond in anchorage zones, or excessive creep and shrinkage of the concrete.

A careful study of the load–deflection curve in Figure 6.1 lead to the following additional observations:

- Due to prestress, the unloaded beam deflects upward, producing camber. Typically, in pretension beds, where the cables are cut from bulkheads, the beam will start deflecting upwards. In order to reduce this deflection, beams will be supported only at the ends, so that self-weight will balance part of the camber. This could be considered as the starting point of the load–deflection curve.
- When the cables are cut, typical age of concrete is only two or three days. Therefore, if proper strength gain is not achieved, excessive camber could occur because the modulus of elasticity of young concrete is low. In some cases, cracks would occur on the top of the beam. Small cracks will

close when the beams are placed in final position and when the external loads are applied. However, excessive cracking should not be permitted, because this might affect the compression behavior when the external loads are applied.

- In typical applications, the beam will still have some camber when the full dead load is applied, and the entire cross-section is in compression.
- When the load–deflection curve crosses the load ($Y$) axis, the stress distribution across the cross-section is uniform compression. This uniform compression could occur only at the maximum bending moment location. The stress distributions at other locations will depend on the type of external load and cable geometry along the beam (see Section 6.3).
- As the load increases, the curve passes through stages in which the tension fiber of concrete reaches a value of zero and, subsequently, the modulus of rupture. When the tension stress reaches the modulus of rupture, the concrete cracks and no longer provides a tension force contribution.
- Further increases in load results in yielding of nonprestressed reinforcement and yielding of prestressed reinforcement, progressively reducing the flexural stiffness.
- In typical cases, excessive strains in tension reinforcement induce crushing of concrete. The maximum load is reached at the onset of crushing of concrete.

The load–deflection response will depend on geometry of cross-section, cable profile along the span, compressive strength of concrete, amount and location of nonprestressed reinforcement, properties, location, area, and magnitude of prestressed reinforcement, time of tensioning (pre- or post-tensioned), and bonding of cables (bonded versus unbonded) cables.

## 6.3   Beam cross-sections and cable profiles

Precast, pretensioned beams are common in the United States of America. Since transportation facilities (which accommodate large trucks and wide roads) allow the transportation of long beams, pretensioned beams are more widely used in the US. Typical cross-sections for parking garages and industrial buildings are double-tee sections (Figure 6.2). Rectangular sections and hollow core sections are used in buildings. For transportation facilities, I and box sections are the popular choices.

The cable profiles could be straight, harped, or draped as shown in Figure 6.3. Theoretically, the profiles can be matched to the bending moment so that when the service load is present, the beam stays horizontal and the entire beam is in uniform compression. However, since the loads cannot be predicted with perfect accuracy and it is neither practical nor economical to produce complex cable profiles, harped tendons are typically used

Actual double-tee sections

| Designation | $b_f$ (in.) | $h_f$ (in.) | $b_{w1}$ (in.) | $b_{w2}$ (in.) | $h$ (in.) | $b$ (in.) |
|---|---|---|---|---|---|---|
| 8DT12 | 96 | 2 | 5.75 | 3.75 | 12 | 48 |
| 8DT14 | 96 | 2 | 5.75 | 3.75 | 14 | 48 |
| 8DT16 | 96 | 2 | 5.75 | 3.75 | 16 | 48 |
| 8DT18 | 96 | 2 | 5.75 | 3.75 | 18 | 48 |
| 8DT20 | 96 | 2 | 5.75 | 3.75 | 20 | 48 |
| 8DT24 | 96 | 2 | 5.75 | 3.75 | 24 | 48 |
| 8DT32 | 96 | 2 | 7.75 | 4.75 | 32 | 48 |
| 10DT32 | 120 | 2 | 7.75 | 4.75 | 32 | 60 |
| 12DT34 | 144 | 4 | 7.75 | 4.75 | 34 | 60 |
| 15DT34 | 180 | 4 | 7.75 | 4.75 | 34 | 60 |

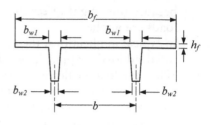

Actual I sections

| Designation | $b_f$ (in.) | $x_1$ (in.) | $x_2$ (in.) | $b_2$ (in.) | $x_3$ (in.) | $x_4$ (in.) | $b_w$ (in.) | $h$ (in.) |
|---|---|---|---|---|---|---|---|---|
| AASHTO 1 | 12 | 4 | 3 | 16 | 5 | 5 | 6 | 28 |
| AASHTO 2 | 12 | 6 | 3 | 18 | 6 | 6 | 6 | 36 |
| AASHTO 3 | 16 | 7 | 4.5 | 22 | 7.5 | 7 | 7 | 45 |
| AASHTO 4 | 20 | 8 | 6 | 26 | 9 | 8 | 8 | 54 |
| AASHTO 5 | 42 | 5 | 7 | 28 | 10 | 8 | 8 | 63 |
| AASHTO 6 | 42 | 5 | 7 | 28 | 10 | 8 | 8 | 72 |

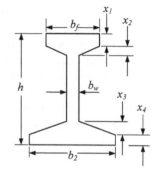

*Figure 6.2* Typical cross-sections for pretensioned beams (Source: Precast/Prestressed Concrete Institute, 1999. Nawy, Edward G., *Prestressed Concrete: A Fundamental Approach*, 4th edn, © 2003, p. 117. Reprinted by permission of Pearson Education, Inc., Upper Saddle River, NJ).

for pretensioned beams. Either harped or parabolic profiles are used for post-tensioned beams.

## 6.4   Stress levels in prestressed tendons (cables) and loss of prestress

The stress–strain behavior of prestressing tendons (cables) is linearly elastic in the initial stage and parabolic (nonlinear) in the postyielding stage. Fracture strengths of tendons are typically 250,000 or 270,000 psi, with initial elastic modulus of $28 \times 10^6$ psi. In most cases, the tendons experience the highest level of stress during the initial prestress. This stress level occurs when they are tensioned between the bulkheads in a pretension bed or against the beams or slabs in post-tensioning. This level of prestress, called "initial prestress," $f_{pi}$, decreases when the tendons (cables) are cut or anchored against beams.

In pretensioned beams, the loss of prestress occurs due to:

- initial elastic shortening of beams;
- creep and shrinkage of concrete;
- relaxation of tendons.

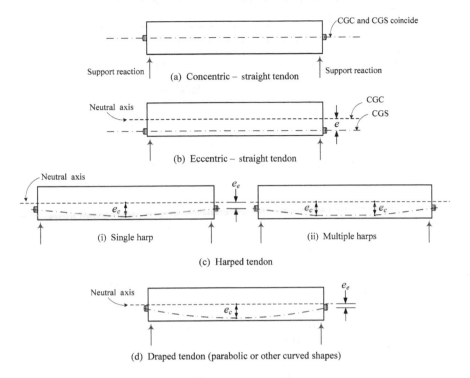

*Figure 6.3* Typical cable profiles.

In post-tensioned beams, the loss occurs due to:

- anchorage seating loss;
- creep and shrinkage of concrete;
- relaxation of tendons.

In addition, the stress may not be the same for the entire length of the tendon due to friction between the tendons and the concrete or the seating used for the tendons.

Creep and shrinkage losses depend on the type of concrete and the age at prestress transfer, while relaxation loss depends on the type of tendons; low relaxation strands have fewer losses. Extensive investigations have been carried out to estimate the losses. Details can be found in textbooks on prestressed concrete (Collins and Mitchell, 2005; Lin and Burns, 1981; Nawy, 2003; Nilson, 1987). For the discussion presented in this book, it is assumed that these losses are available from the construction documents. As an approximation, losses can be estimated using guidelines provided by the American Association of State Highway and Transportation Officials (AASHTO), Table 6.1. Losses due to friction are not included. These losses should be calculated according to Section 6.5 of the AASHTO specifications

*Table 6.1* Lump-sum losses recommended by AASHTO (1975) and Post-Tensioning Institute (1999)

| AASHTO lump-sum losses | | |
|---|---|---|
| *Type of prestressing steel* | *Total loss* | |
| | $f_c' = 4000\,psi$ | $f_c' = 5000\,psi$ |
| Pretensioning strand | | 45 ksi |
| Postensioning wire or strand | 32 ksi | 33 ksi |
| Bars | 22 ksi | 23 ksi |
| Approximate prestress loss values for post-tensioning | | |
| *Tendon type* | *Prestress loss* | |
| | *Slabs* | *Beams and joists* |
| Stress-relieved 270 K strand and stress-relieved 240 K wire | 30 ksi | 35 ksi |
| Bar | 20 ksi | 25 ksi |
| Low-relaxation 270 K strand | 15 ksi | 20 ksi |

(1975). The stress after the loss of prestress, and when the weight of the beam is acting, is designated as effective prestress, $f_{pe}$.

Since the stresses have to be checked at various stages, the following definitions are provided for clarity:

- initial prestress, $f_{pi}$;
- total prestressed losses, $\Delta f_{pt}$;
- effective prestress, $f_{pe}$;
- yield stress of tendons, $f_{py}$;
- ultimate (fracture) stress of tendons, $f_{pu}$;
- tendon stress at any given stage of loading, $f_{ps}$.

## 6.5   Analysis of unstrengthened prestressed concrete beams: general discussion

The analysis has to be carried out at:

- initial prestressing;
- when service loads are present;
- at failure.

In terms of cross-sectional behavior, the critical levels are:

- uncracked section;
- cracking moment capacity;

- properties of cracked section;
- moment capacity when nonprestressed steel yields;
- moment capacity when prestressed steel yields;
- ultimate moment capacity, $\phi M_n$.

The critical locations may not be maximum-bending-moment locations and, therefore, analyses may have to be carried out at various locations along the beam. For example, stresses at supports where the eccentricity is minimum, could control the magnitude of the prestress.

The beams could be partially or fully prestressed with bonded or unbonded tendons. Partially prestressed concrete beams contain both prestressed and nonprestressed steel; partially prestressed beams are chosen for discussion within this chapter. If nonprestressed steel is not present, their area can be taken as zero. Both bonded and unbonded cables are discussed.

For clarity, the analysis is presented in separate sections at various stages of loading. Again, only the summary is presented and the reader should consult the textbooks on prestressed concrete for a more thorough discussion (Collins and Mitchell, 2005; Lin and Burns, 1981; Nawy, 2003; Nilson, 1987).

## 6.6 Uncracked section analysis: unstrengthened beams

It is assumed that the geometric properties of the cross-section, reinforcement details, and level of prestress are known. For standard sections, section properties such as moment of inertia can be found in Prestressed Concrete Institute (PCI) (1999) or AASHTO Handbooks (1975).

The basic principles of mechanics are applied to solve for stresses and deformations. The equations are same as those used for reinforced concrete. However, the presence of prestress at an eccentricity, $e$, induces axial and bending stresses. The change in curvature and strains across the thickness are not significant. However, for beams with bonded tendons, the stress in the prestressed tendon does change. For unbonded tendons, the change in prestress is beam dependent and, hence, the increase can be neglected. Therefore, for beams with unbonded tendons, the prestress $f_{ps}$ can be assumed to be the effective prestress, $f_{pe}$, for computation of stresses, strains, and deflection. The mechanism for increase in prestress is further explained in the next section.

When the self-weight of the beam is the only load, creating a maximum moment of $M_D$, stress in the prestressed tendon is:

$$f_{ps} = f_{pe} \qquad (6.1)$$

$$\varepsilon_{pe} = \frac{f_{pe}}{E_{ps}} \qquad (6.2)$$

where $E_{ps}$ is the modulus of elasticity of prestressed tendons. Concrete stress at the level of prestressed tendon:

$$\varepsilon_{ce} = \frac{1}{E_c}\left[\frac{(A_{ps}f_{pe}e_0 - M_D)}{I_g}e_0 + \frac{A_{ps}f_{pe}}{A_g}\right]$$

(6.3)

where:

$A_{ps}$ is the area of prestressed steel
$E_c$ is the modulus of elasticity of concrete
$I_g$ and $A_g$ are gross moment of inertia and gross area of cross-section, respectively
$e_0$ is the maximum eccentricity

If the external moment, $M$ is less than the cracking moment, $M_{cr}$:
For beams with unbonded tendons:

$$f_{ps} = f_{pe}$$

(6.4)

For beams with bonded tendons:

$$f_{ps} = f_{pe} + \frac{(M - M_D)e_0}{I_g\frac{E_c}{E_{ps}} + A_{ps}\left(\frac{I_g}{A_g} + e_0^2\right)}$$

(6.5)

Maximum stress in concrete (compression):

$$f_{ct} = \frac{(M - A_{ps}f_{ps}e_0)}{I_g}y_t + \frac{A_{ps}f_{ps}}{A_g}$$

(6.6)

where $y_t$ is the distance of the top fiber from the neutral axis, which is the same as the depth of the neutral axis of the uncracked section, $\overline{y}$. Stress in nonprestressed steel:

$$f_s = \frac{E_s}{E_c}\left[-\left(\frac{M - A_{ps}f_{ps}e_0}{I_g}\right)(d_s - y_t) + \frac{A_{ps}f_{ps}}{A_g}\right]$$

(6.7)

where $d_s$ is the effective depth of nonprestressed steel. Note that a negative number for $f_s$ represents tension. $A_{ps}f_{ps}$ is the prestressing force that produces uniform compression, and $A_{ps}f_{ps}e_0$ is the moment generated by the prestress that counteracts the external moment or produces tension in the top fiber. Stress at the bottom fiber:

$$f_b = -\left(\frac{M - A_{ps}f_{ps}e_0}{I_g}\right)y_b + \frac{A_{ps}f_{ps}}{A_g}$$

(6.8)

When this stress reaches the value of the modulus of rupture, $f_r$, the section cracks. Therefore, the moment capacity at cracking can be computed using Equation (6.8) by replacing $f_b$ with $f_r$. For beams with unbonded tendons $f_{ps} = f_{pe}$, whereas for beams with bonded tendons, $f_{ps}$ is a function of $M_{cr}$. Using the proper values for $f_{ps}$:

For beams with unbonded tendons:

$$M_{cr} = \frac{I_g}{y_b}f_r + A_{ps}f_{pe}e_0 + \frac{A_{ps}f_{pe}}{A_g}\frac{I_g}{y_b} \qquad (6.9)$$

For beams with bonded tendons:

$$M_{cr} = \frac{-M_D e_0 A_{ps}\left(e_0 + \frac{I_g}{A_g y_b}\right) + \left[I\frac{E_c}{E_{ps}} + A_{ps}\left(\frac{I_g}{A_g} + e_0^2\right)\right]\left[(A_{ps}f_{se})\left(e_0 + \frac{I_g}{A_g y_b}\right) + 7.5\sqrt{f_c'}\frac{I_g}{y_b}\right]}{I_g\frac{E_c}{E_{ps}} + \frac{A_{ps}I_g}{A_g}\left(1 - \frac{e_0}{y_b}\right)}$$

$$(6.10)$$

The deflection of the uncracked beam can be computed using $I_g$ and the modulus of elasticity of concrete, $E_c$. The effect of prestress should be considered in deflection calculations. Typically, the prestress is converted to equivalent moment or load. For example, if the tendon eccentricity $e_o$ is the same for the entire length, the effect of prestress is equivalent to constant moment, $A_{ps}f_{ps}e_o$. Equations for typical external loads and prestressing tendon profiles are shown in Figure 6.4.

## 6.7 Computation of cracking moment and deflections

### Example 6.1: Rectangular cross-section with straight tendon

For the rectangular beam shown in Figure 6.5, compute maximum stresses and deflection for uniform load of 500 lb/ft. Also, compute the cracking moment. Compressive strength of concrete, $f_c' = 60,000$ psi. Nonprestressed steel, $A_s$, consists of two ASTM #8 bars. The prestressed reinforcement consists of two cables with a nominal diameter of 0.5 in. with area, $A_{ps} = 0.306$ in.$^2$ Effective prestress, $f_{pe} = 150,000$ psi. Assume $E_s = 29 \times 10^6$ psi, $E_c = 4.3 \times 10^6$ psi, $E_{ps} = 27 \times 10^6$ psi, and assume unbonded tendons. (Bonded tendons are presented in Example 6.2.)

Solution:

(a) Unbonded tendons
   Depth of neutral axis, $\overline{y} = 6$ in.

$$y_t - y_b = 6 \text{ in.}$$

$$I_g = \frac{bh^3}{12} = \frac{12 \times 24^3}{12} = 13,824 \text{ in.}^4$$

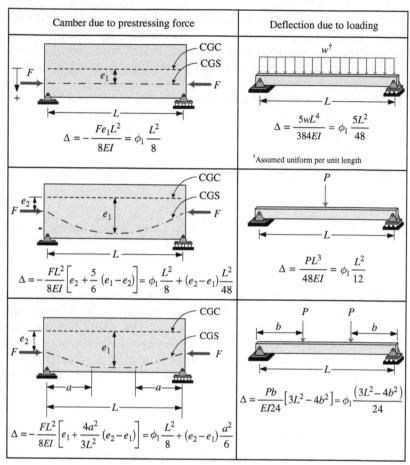

| Camber due to prestressing force | Deflection due to loading |
|---|---|
| CGC, CGS, $F$, $e_1$, $L$ <br> $\Delta = -\dfrac{Fe_1 L^2}{8EI} = \phi_1 \dfrac{L^2}{8}$ | $w^\dagger$, $L$ <br> $\Delta = \dfrac{5wL^4}{384EI} = \phi_1 \dfrac{5L^2}{48}$ <br><br> †Assumed uniform per unit length |
| CGC, CGS, $F$, $e_2$, $e_1$, $L$ <br> $\Delta = -\dfrac{FL^2}{8EI}\left[e_2 + \dfrac{5}{6}(e_1 - e_2)\right] = \phi_1 \dfrac{L^2}{8} + (e_2 - e_1)\dfrac{L^2}{48}$ | $P$, $L$ <br> $\Delta = \dfrac{PL^3}{48EI} = \phi_1 \dfrac{L^2}{12}$ |
| CGC, CGS, $F$, $e_2$, $e_1$, $a$, $L$ <br> $\Delta = -\dfrac{FL^2}{8EI}\left[e_1 + \dfrac{4a^2}{3L^2}(e_2 - e_1)\right] = \phi_1 \dfrac{L^2}{8} + (e_2 - e_1)\dfrac{a^2}{6}$ | $P$, $P$, $b$, $b$, $L$ <br> $\Delta = \dfrac{Pb}{EI24}\left[3L^2 - 4b^2\right] = \phi_1 \dfrac{\left(3L^2 - 4b^2\right)}{24}$ |

*Figure 6.4* Typical mid-span deflections for simply supported beams.
*Notes:* Subscript "1" denotes mid-span; subscript "2" denotes support. For the uncracked section, use $I$ of transformed section or $I_{\text{gross}}$ as first approximation. For the cracked section, use $I_e$ = effective moment of inertia.

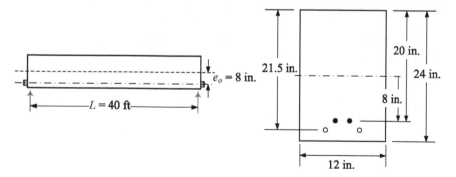

$e_o = 8$ in.    $L = 40$ ft    20 in.    21.5 in.    24 in.    8 in.    12 in.

*Figure 6.5* Details of beam for Example 6.1.

Maximum moment, $M = \dfrac{qL^2}{8} = \dfrac{500 \times 40^2}{8} \times 12 = 1{,}200{,}000 \text{ in.-lb}$

$f_{ps} = f_{pe} = 150{,}000 \text{ psi}$

Maximum stress in concrete:

$$f_{ct} = \frac{(1{,}200{,}000 - 0.306 \times 150{,}000 \times 8)}{13{,}824} \times 12 + \frac{0.306 \times 150{,}000}{12 \times 24}$$

$f_{ct} = 723 + 160 = 833 \text{ psi, compression}$

Stress in nonprestressed steel:

$$f_s = \frac{29}{4.3}\left[\frac{-(1{,}200{,}000 - 0.306 \times 150{,}000 \times 8)}{13{,}824} \times 9.5\right.$$
$$\left. +\frac{0.306 \times 150{,}000)}{12 \times 24}\right]$$

$f_s = 2781 \text{ psi, tension}$

Stress in bottom fiber:

$$f_b = \frac{-(1{,}200{,}000 - 0.306 \times 150{,}000 \times 8)}{13{,}824} \times 12 + \frac{0.306 \times 150{,}000}{12 \times 24}$$

$f_b = -723 + 160 = -563 \text{ psi (tension)}$

$< 7.5\sqrt{6000} = 581 \text{ psi}$

Therefore, no cracking.

Computation of deflection:

$I_g = 13{,}824 \text{ in.}^4$

$$\delta = \frac{5(500)(40)^4(12)^3}{384 \times 4.3 \times 10^6 \times 13{,}824} - \frac{(A_{ps}f_{se})(40)^2(12)^3}{8 \times 4.3 \times 10^6 \times 13{,}824}$$

$= 0.485 - 0.022 = 0.47 \text{ in.}$

$f_r = 7.5\sqrt{6000} = 581 \text{ psi}$

Computation of cracking moment using Equation (6.9):

$$M_{cr} = \frac{13{,}824}{12} \times 581 + 0.306 \times 150{,}000 \times 8 + \frac{0.306 \times 150{,}000}{12 \times 24} \times \frac{13{,}824}{12}$$

$M_{cr} = 1{,}220{,}110 \text{ in.-lb}$

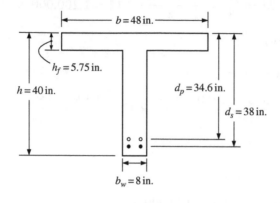

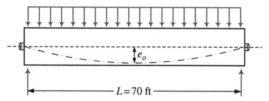

*Figure 6.6* Details of beam for Example 6.2.

Note that the compression induced at the bottom fiber by prestress increases the cracking moment considerably.

### Example 6.2: T-beam with parabolic tendon

Compute the stresses for a maximum moment of 5,505,560 in.-lb and the cracking moment for the T-beam shown in Figure 6.6. Area of nonprestressed reinforcement, $A_s = 1.8$ in.$^2$ (3 # 7 bars). Area of prestressed reinforcement, $A_{ps} = 1.07$ in.$^2$ (7 – 1/2 in. strands). Cable profile: parabolic with zero end eccentricity. Compressive strength of concrete is 5000 psi. Modulus of elasticity, $E_c = 4.3 \times 10^6$ psi, $E_s = 29 \times 10^6$ psi, and $E_{ps} = 29 \times 10^6$ psi. Moment due to self-weight, $M_D = 4,505,560$ in.-lb. Effective prestress, $f_{pe} = 150,000$ psi.

(a)  Assume unbonded tendons
(b)  Assume bonded tendons

Solution:

(a) Unbonded tendons

Depth of neutral axis for uncracked section, $A\overline{Y} = \sum_i^n A_i y_i$ (Figure 6.7).

$$\overline{Y} = \frac{48 \times \dfrac{5.75^2}{2} + 8 \times (40 - 5.75)\left(5.75 + \dfrac{(40 - 5.75)}{2}\right)}{48 \times 5.75 + 8(40 - 5.75)} = 12.9 \text{ in.}$$

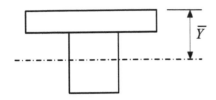

*Figure 6.7* Depth of neutral axis for uncracked section for Example 6.2.

Therefore, $y_t = 12.9$ in., $y_b = 40 - 12.9 = 27.1$ in.
Maximum eccentricity, $e_o = d_p - y_t = 34.6 - 12.9 = 21.7$ in.

$$I_g = \frac{b\bar{y}^3}{3} + \frac{(b - b_w)(\bar{y} - h_f)^3}{3} + \frac{b_w(h - \bar{y})^3}{3}$$

$$I_g = 82,065 \text{ in.}^4$$

$$A_g = 48 \times 5.75 + (40 - 5.75) = 550 \text{ in.}^2$$

$$f_{ps} = f_{pe} = 150,000 \text{ psi}$$

Using Equation (6.6):

$$f_{ct} = \frac{(5,505,500 - 1.07 \times 150,000 \times 21.7)}{82,065} \times 12.9$$

$$+ \frac{1.07 \times 150,000}{550} = 610 \text{ psi}$$

Using Equation (6.7):

$$f_s = \frac{29}{4.3} \left[ - \left( \frac{\frac{(5,505,560 - 1.07 \times 150,000 \times 21.7)}{82,065}}{\times (38 - 12.9) + \frac{1.07 \times 150,000}{550}} \right) \right] = 2200 \text{ psi}$$

$$y_b = (h - \bar{y}) = 40 - 12.9 = 27.1 \text{ in.}$$

$$f_r = 7.5\sqrt{5000} = 530 \text{ psi}$$

Computation of cracking moment using Equation (6.9):

$$M_{cr} = \frac{82,065}{27.1} \times 530 + 1.07 \times 150,000 \times 21.7 + \frac{1.07 \times 150,000}{550}$$

$$\times \frac{82,065}{27.1} = 5,971,503$$

$$M_{cr} = 5,971,503 \text{ in.-lb}$$

(b) Bonded tendons

$$A_g = 550 \text{ in.}^2$$

$$I_g = 82,065 \text{ in.}^4$$

$$y_t = 12.9 \text{ in.}$$

$$y_b = 27.1 \text{ in.}$$

$$e_o = 21.7 \text{ in.}$$

$$f_r = 530 \text{ psi}$$

Using Equation (6.5), stress in prestressed steel:

$$f_{ps} = 150,000 + \frac{(5,505,560 - 4,505,560) \times 21.7}{82,065 \left(\dfrac{4.3}{27}\right) + 1.07 \left(\dfrac{82,065}{550} + 21.7^2\right)}$$

$$f_{ps} = 151,580 \text{ psi (Therefore, stress increase} = 1580 \text{ psi)}$$

Using Equation (6.6), maximum concrete stress:

$$f_{ct} = \frac{(5,505,560 - 1.07 \times 151,580 \times 21.7)}{82,065} \times 12.9 + \frac{1.07 \times 151,580}{550} = 607 \text{ psi}$$

Using Equation (6.7), stress in nonprestressed steel:

$$f_s = \frac{29}{4.3} \left[ - \left( \frac{\dfrac{(5,505,560 - 1.07 \times 151,580 \times 21.7)}{82,065}}{\times (38 - 12.9) + \dfrac{1.07 \times 151,580}{550}} \right) \right] \text{psi}$$

$$f_s = 2108 \text{ psi}$$

Using Equation (6.10), cracking moment capacity:

$$M_{cr} = \frac{-4,505,560 \times 21.7 \times 1.07 \left(21.7 + \dfrac{82,065}{27.1 \times 550}\right) + \left[ \begin{matrix} \left(82,065 \times \dfrac{4.3}{27}\right) + 1.07 \left(21.7 + \dfrac{82,065}{27.1 \times 550} + 21.7^2\right) \times \\ \left[1.07 \times 150,000 \times \left(21.7 + \dfrac{82,065}{27.1 \times 550}\right) + 7.5 \sqrt{5000} \dfrac{82,065}{27.1}\right] \end{matrix} \right]}{\left(82,605 \times \dfrac{4.3}{27}\right) + \left(\dfrac{1.07 \times 82,065}{550}\right) \left(1 - \dfrac{21.7}{27.1}\right)}$$

$$M_{cr} = \frac{(-123 \times 10^6) + (13{,}733 \times 5.9 \times 10^6)}{13{,}101}$$

$$M_{cr} = 6{,}251{,}232 \text{ in.-lb}$$

The 4.5% increase in cracking moment over unbonded tendons is due to increase in stress in prestressed steel. As the curvature changes, the strain at the level of prestress increases, resulting in an increase of prestress. In some cases, the more elaborate analysis to obtain this accuracy may not be warranted, especially since effective stress and modulus of rupture of concrete are approximate. However, since most computations are performed using computers, use of extensive equations may not require considerable additional time. The judgment to consider or neglect the prestress increase in bonded tendons for uncracked sections is left to the user.

## 6.8  Cracked section analysis: unstrengthened beams

Equations for cracked section analysis can be developed using strain compatibility, force, and moment equilibrium equations. Since these equations are available in the literature, their use is illustrated without elaborating the derivations. The basic differences between the reinforced and prestressed concrete beams are the contribution of prestress and the change in prestress with increase in moment. The change in prestress is different for bonded and unbonded tendons. Since in bonded tendons, the change in prestress, $\Delta f_{ps}$, is section dependent, its magnitude is significant, whereas in the case of unbonded tendons, the change, $\Delta f_{ps}$, is beam dependent and the magnitude is not very significant. Therefore, these two groups of beams are discussed separately.

Typical variation of increase in prestress is presented in Figure 6.8. The prestress value is designated as effective prestress, $f_{pe}$, when the moment due to the self-weight of the beam is the only external moment. When additional load is added, the stress increases because the strain at the level of prestress increases. Significant changes occur when the beam cracks, nonprestressed steel yields, and prestressed steel yields.

### 6.8.1  Beams with bonded tendons

For prestressed concrete beams, the equation for the depth of the neutral axis for the cracked section is obtained using force equilibrium. The compression force generated by concrete and steel in the compression zone should be equal to the force generated by tension steel that includes nonprestressed and prestressed steel. Due to the presence of prestress, the equation is also a function of external moment $M$. This moment should be greater than the cracking

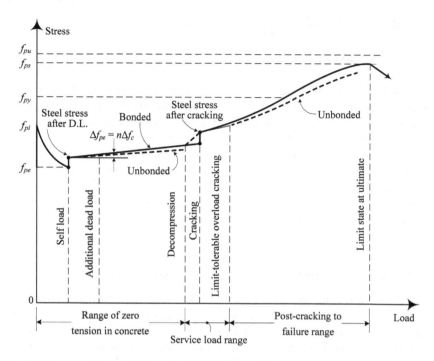

*Figure 6.8* Prestressing steel at various load levels (Source: Nawy, Edward G., *Prestressed Concrete: A Fundamental Approach*, 4th edn, © 2003, p. 179. Reprinted by permission of Pearson Education, Inc., Upper Saddle River, NJ).

moment, $M_{cr}$. Concrete and the reinforcement should be in the elastic range. The cubic equation can be written in the form (Naaman, 2004):

$$\frac{\lambda_1}{M}(kd)^3 + \left(b_w - \frac{\lambda_2}{M}\right)(kd)^2 + \left(\lambda_3 - \frac{\lambda_4}{M}\right)(kd) - \left(\lambda_5 + \frac{\lambda_6}{M}\right) = 0 \quad (6.11)$$

where $kd$ is the depth of neutral axis and $\lambda_i$ are constants.

$$\lambda_1 = \frac{A_{ps}E_{ps}}{3}(\varepsilon_{pe} + \varepsilon_{ce})b_w \quad (6.12)$$

where $\varepsilon_{ce}$ is the maximum concrete strain when only self-weight moment, $M_d$, is present, Equation (6.3):

$$\lambda_2 = A_{ps}E_{ps}(\varepsilon_{pe} + \varepsilon_{ce})b_w d_p \quad (6.13)$$

$$\lambda_3 = 2(b - b_w)h_f + 2\frac{A_s E_s}{E_c} + 2\frac{A'_s E'_s}{E_c} + 2\frac{A_{ps}E_{ps}}{E_c} \quad (6.14)$$

$$\lambda_4 = A_{ps}E_{ps}(\varepsilon_{pe} + \varepsilon_{ce})\left[2(b - b_w)h_f d_p - (b - b_w)h_f^2 - 2\frac{A_s E_s}{E_c}(d_s - d_p)\right.$$

$$\left. - 2\frac{A_s' E_s'}{E_c}(d_s' - d_p)\right] \tag{6.15}$$

$$\lambda_5 = (b - b_w)h_f^2 + \frac{2}{E_c}(A_s E_s d_s + A_s' E_s' + A_{ps} E_{ps} d_p) \tag{6.16}$$

$$\lambda_6 = A_{ps}E_{ps}(\varepsilon_{pe} + \varepsilon_{ce})\left[\frac{2}{3}(b - b_w)h_f^3 + 2\frac{A_s E_s}{E_c}(d_s - d_p)d_s\right.$$

$$\left. + 2\frac{A_s' E_s'}{E_c}(d_s' - d_p) - (b - b_w)h_f^2 d_p\right] \tag{6.17}$$

Once the depth of the neutral axis is known, the maximum concrete stress is:

$$f_{ct} = \frac{A_{ps}E_{ps}(\varepsilon_{pe} + \varepsilon_{ce})(kd)}{\dfrac{b(kd)^2}{2} - \dfrac{(b - b_w)(c - h_f)^2}{2} - \dfrac{A_{ps}E_{ps}}{E_c}(d_p - kd) - \dfrac{A_s' E_s'}{E_c}(d_s' - kd)} \tag{6.18}$$

$$f_s = \frac{E_s}{E_c}f_{ct}\left(\frac{d_s - kd}{kd}\right) \tag{6.19}$$

$$f_s' = \frac{E_s'}{E_c}f_{ct}\left(\frac{kd - d'}{kd}\right) \tag{6.20}$$

$$f_{ps} = E_{ps}(\varepsilon_{pe} + \varepsilon_{ce}) + \frac{E_{ps}}{E_c}f_{ct}\left(\frac{d_p - kd}{kd}\right) \tag{6.21}$$

Moment of inertia of the cracked section:

$$I_{cr} = \frac{b(kd)^3}{3} - \frac{(b - b_w)(kd - h_f)^3}{3} + \frac{A_{ps}E_{ps}}{E_c}(d_p - kd)^2 + \frac{A_s E_s}{E_c}(d_s - kd)^2$$

$$+ A_s'\left(\frac{E_s'}{E_c} - 1\right)(kd - d_s')^2 \tag{6.22}$$

Effective moment of inertia:

$$I_e = I_{cr} + (I_g - I_{cr})\left(\frac{M_{cr}}{M}\right)^3 \le I_g \tag{6.23}$$

These equations are very general and cover the entire range from reinforced concrete to fully prestressed concrete with or without compression reinforcement. For rectangular sections, use $b = b_w$. Use of these equations is illustrated in numerical examples presented in Section 6.9.

### 6.8.2   *Beams with unbonded tendons*

As mentioned earlier, the difference between bonded and unbonded tendons is the change in stress level of prestressed reinforcement. In both cases, the stress in prestressed steel under the weight of the beam is taken as effective prestress, $f_{pe}$. The stress level in the unbonded tendon under moments greater than the cracking moment, is greater than $f_{pe}$ and less than $f_{ps}$ calculated using Equation (6.18). If there is a considerable amount of friction between the tendon and the ducts, the stress will be closer to the value calculated using Equation 6.18. However, there is a good chance of slippage, especially due to repeated loading. If there is considerable slippage, the increase in stress depends on the elongation of the beam at the tendon level due to deflection. This elongation is not significant until nonprestressed or prestressed steel yields and creates a hinge (Balaguru, 1981). Therefore, it is recommended to neglect the stress increase in the tendons.

This assumption will result in overestimation of stresses in concrete and nonprestressed reinforcement and underestimation of stress increase in tendons. If it is suspected that stresses in prestressed tendon could reach critical level, the following equations can be used to estimate the increase in prestress:

$$\Delta f_{ps} = E_{ps} \Delta \varepsilon_{ps} \tag{6.24}$$

$$\Delta \varepsilon_{ps} = \alpha_1 \left( \frac{\delta}{e} \right) - \alpha_2 \left( \frac{\delta}{e} \right)^2 \tag{6.25}$$

where $\delta$ = maximum deflection, $e$ = maximum eccentricity:

$$\alpha_1 = 0.000923 \left( \frac{e}{L} \right) + 5.11 \left( \frac{e}{L} \right)^2 \quad \text{for} \quad 0 < \left( \frac{e}{L} \right) < 0.06 \tag{6.26}$$

$$\alpha_2 = \left( \frac{11}{107} \right) e^{(135 \frac{e}{L})} \text{ for } 0 < \left( \frac{e}{L} \right) \leq 0.04 \tag{6.27}$$

$$\alpha_2 = \left( \frac{1}{105} \right) e^{(75 \frac{e}{L})} \text{ for } 0.04 < \left( \frac{e}{L} \right) < 0.06$$

The equations for beams with unbonded tendons are similar to equations for bonded tendons presented in Section 6.8.1. The modified equations account for zero contribution of prestressed steel due to curvature change at the cross-section. For simplicity and comparison, the equations are presented in a similar form. $E_{ps}\varepsilon_{pe}$ is changed to $f_{pe}$ since the effect of $\varepsilon_{ce}$ is not present for unbonded beams.

The equation for the depth of the neutral axis is:

$$\frac{\lambda_1''}{M}(kd)^3 + \left( b_w - \frac{\lambda_2''}{M} \right)(kd)^2 + \left( \lambda_3'' - \frac{\lambda_4''}{M} \right)(kd) - \left( \lambda_5'' + \frac{\lambda_6''}{M} \right) = 0 \tag{6.28}$$

$$\lambda_1^u = \frac{A_{ps}f_{pe}}{3}b_w \tag{6.29}$$

$$\lambda_2^u = A_{ps}f_{pe}b_w d_p \tag{6.30}$$

$$\lambda_3^u = 2(b - b_w)h_f + 2\left(\frac{A_s E_s}{E_c}\right) + 2\left(\frac{A_s' E_s'}{E_c}\right) \tag{6.31}$$

$$\lambda_4^u = A_{ps}f_{pe}\left[2(b - b_w)h_f d_p - (b - b_w)h_f^2 - 2\left(\frac{A_s E_s}{E_c}\right)(d_s - d_p)\right.$$

$$\left. -2\left(\frac{A_s' E_s'}{E_c}\right)(d_s' - d_p)\right] \tag{6.32}$$

$$\lambda_5^u = (b - b_w)h_f^2 + \left(\frac{2A_s E_s d_s}{E_c}\right) + \left(\frac{2A_s' E_s' d_s'}{E_c'}\right) \tag{6.33}$$

$$\lambda_6^u = A_{ps}f_{pe}\left[\left(\frac{2}{3}\right)(b - b_w)h_f^3 + 2\frac{A_s E_s}{E_c}(d_s - d_p)d_s + 2\frac{A_s' E_s'}{E_c'}(d_s' - d_p)d_s'\right.$$

$$\left. -(b - b_w)h_f^2 d_p)\right] \tag{6.34}$$

$$f_{ct} = \frac{A_{ps}f_{pe}(kd)}{\dfrac{b(kd)^2}{2} - \dfrac{(b - b_w)(kd - h_f)^2}{2} - \dfrac{A_s E_s}{E_c}(d_s - kd) - \dfrac{A_s' E_s'}{E_c}(d_s' - kd)} \tag{6.35}$$

$$f_{ps} = f_{pe} \tag{6.36}$$

$$f_s = \frac{E_s}{E_c}f_{ct}\left(\frac{d_s - kd}{kd}\right) \tag{6.37}$$

$$f_s' = \frac{E_s'}{E_c}f_{ct}\left(\frac{kd - d'}{kd}\right) \tag{6.38}$$

Moment of inertia of the cracked section:

$$I_{cr} = \frac{b(kd)^3}{3} - \frac{(b - b_w)(kd - h_f)^3}{3} + \frac{A_s E_s}{E_c}(d_s - kd)^2 + \frac{A_s' E_s'}{E_c}(kd - d')^2 \tag{6.39}$$

The equation for the effective moment of inertia is the same as Equation (6.23).

## 6.9   Computation of stresses and deflection for working loads: unstrengthened beams

*Example 6.3: Bonded tendons*
For the T-beam of Example 6.2, compute the deflection and stresses at mid-span assuming bonded tendons. The maximum moment is 7,445,560 in.-lb. Simply supported span is 70 ft. Assume $E_s = 29 \times 10^6$ psi, $E_{ps} = 27 \times 10^6$ psi, $E_c = 4.3 \times 10^6$ psi, $M_D = 4,505,560$ in.-lb, and cracking moment, $M_{cr} = 6,251,232$ in.-lb.

Solution:

$$A_s = 1.8 \text{ in.}^2$$

$$A_{ps} = 1.07 \text{ in.}^2$$

$$A_g = 550 \text{ in.}^2$$

$$I_g = 82,065 \text{ in.}^4$$

$$\overline{y} = y_t = 12.9 \text{ in.}$$

$$y_b = 27.1 \text{ in.}$$

$$\varepsilon_{pe} = \frac{150,000}{29 \times 10^6} = 0.0052 \text{ in./in.}$$

$$M_{cr} = 6,251,232 \text{ in.-lb}$$

Compute the depth of the neutral axis:

$$\varepsilon_{ce} = \frac{1}{4.3 \times 10^6}\left[\frac{1.07 \times 150,000}{82,065}(149.2 + 21.7)\right.$$

$$\left. + \frac{4,505,560}{82,065} \times 21.7\right]$$

$$\varepsilon_{ce} = 0.00036 \text{ in./in.}$$

$$\varepsilon_{ce} + \varepsilon_{pe} = 0.00556 \text{ in./in.}$$

$$\left(\frac{\lambda_1}{M}\right) = \frac{1.07 \times 27 \times 10^6(0.00556) \times 8}{3 \times 7,445,560} = 0.057(\lambda_1 = 428,342)$$

$$\left(b_w - \frac{\lambda_2}{M}\right) = \left[8 - \frac{1.07 \times 27 \times 10^6(0.00556)8 \times 34.6}{7,445,560}\right]$$

$$= 1.985(\lambda_1 = 44,785,043)$$

$$\left(\lambda_3 - \frac{\lambda_4}{M}\right) = \begin{bmatrix} 2 \times (48-8) \times 5.75 + \dfrac{2 \times 1.8 \times 29 \times 10^6}{4.3 \times 10^6} + 0 \\[4mm] + \dfrac{2 \times 1.07 \times 27 \times 10^6}{4.3 \times 10^6} \end{bmatrix}$$

$$- \frac{1.07 \times 27 \times 10^6 \times 0.00556}{7,445,560}$$

$$\begin{bmatrix} 2(48-8) \times 5.75 \times 34.6 - (48.8)5.75^2 - \\[2mm] \dfrac{2 \times 1.8 \times 29 \times 10^6}{4.3 \times 10^6}(38-34.6) - 0 \end{bmatrix}$$

$$= 497.8 - 0.0217 \times 14,570 = 182.92$$

$(\lambda_3 = 497.8, \quad \lambda_4 = 2,344,361,141)$

$$\left(\lambda_5 + \frac{\lambda_6}{M}\right)$$

$$= \left[(48-8) \times 5.75^2 + \frac{2 \times 10^6}{4.3 \times 10^6}(1.8 \times 29 \times 38 + 0 + 1.07 \right.$$

$$\left. \times 27 \times 34.6)\right] + \frac{1.07 \times 27 \times 10^6 \times 0.00556}{7,445,560}$$

$$\begin{bmatrix} \dfrac{2}{3}(48-8) \times 5.75^3 + \dfrac{2 \times 1.8 \times 29 \times 10^6}{4.3 \times 10^6}(38-34.6) \\[2mm] \times 38 + 0 - (48-8) \times 5.75^2 \times 34.6 \end{bmatrix}$$

$$\left(\lambda_5 + \frac{\lambda_6}{M}\right) = 1904.36$$

$(\lambda_5 = -2710, \lambda_6 = -6,031,520,292)$

Therefore, equation for the depth of neutral axis, $kd$, is:

$$0.057(kd)^3 + 1.985(kd)^2 + 182.92(kd) - 1904.36 = 0$$

$$(kd) = 9.25 \text{ in.}$$

Maximum concrete stress:

$$(f_{ct})_{max} = \frac{1.07 \times 27 \times 10^6 (0.00556) \times 9.25}{\left\{ \begin{array}{l} \dfrac{48 \times 9.25^2}{2} - \dfrac{(48-8)}{2}(9.25-5.75)^2 \\[3mm] - \dfrac{1.07 \times 27 \times 10^6}{4.3 \times 10^6}(34.6-9.2) - \dfrac{1.8 \times 29 \times 10^6}{4.3 \times 10^6}(38-9.25) - 0 \end{array} \right\}}$$

$(f_{ct})_{\max} = 1171\,\text{psi}$

$$(f_s)_{\max} = \frac{1171 \times 29 \times 10^6}{4.3 \times 10^6}\left(\frac{38 - 9.25}{9.25}\right)$$

$(f_s)_{\max} = 24{,}970\,\text{psi}$

$$(f_{ps})_{\max} = 27 \times 10^6 \times (0.00556) + \frac{1171 \times 29 \times 10^6}{4.3 \times 10^6}\left(\frac{34.6 - 9.25}{9.25}\right)$$

$(f_{ps})_{\max} = 170{,}633\,\text{psi}$

Taking the moment about the center of gravity of the prestressed steel, the moment capacity is 7,436,321in.-lb; error < 0.1%.

$$I_{cr} = \frac{48 \times 9.15^3}{3} - \frac{(48 - 8)(9.15 - 5.75)^3}{3} + 1.07 \times \frac{27}{43}(34.6 - 9.15)^2$$

$$+ 1.8 \times \frac{29}{43}(38 - 9.15)^2$$

$I_{cr} = 26{,}189\,\text{in.}^4$

$$I_e = 26{,}189 + (82{,}065 - 26{,}189)\left(\frac{6{,}251{,}232}{7{,}445{,}560}\right)$$

$I_e = 55{,}023\,\text{in.}^4$

$$\delta_{\max} = \frac{5ML^2}{48E_cI_e} - \frac{(A_{ps}f_{ps})L^2}{8EI} \times \frac{5}{6} \times \text{max eccentricity}$$

$$\delta_{\max} = \frac{5 \times 7{,}445{,}560 \times (70 \times 12)^2}{48 \times 4.3 \times 10^6 \times 55{,}023} - \frac{1.07 \times 170{,}633 \times (70 \times 12)^2}{8 \times 4.3 \times 10^6 \times 55{,}023}$$

$$\times \frac{5}{6} \times 2.17$$

$\delta_{\max} = 2.31 - 1.23 = 1.1\,\text{in.}$

### Example 6.4: Unbonded tendons
Repeat Example 6.3 for unbonded tendons.

Solution:

Again, based on Example 6.2, the following are known:

$A_s = 1.8\,\text{in.}^2$

$A_{ps} = 1.07\,\text{in.}^2$

$e_0 = 21.7\,\text{in.}$

$A_g = 550 \text{ in.}^2$

$I_g = 82,065 \text{ in.}^4$

$y = \bar{y} = 12.9 \text{ in.}$

$y_b = 27.1 \text{ in.}$

$M = 7,445,560 \text{ in.-lb}$

$M_{cr} = 5,972,054 \text{ in.-lb}$

$\varepsilon_{pe} = 0.0052 \text{ in./in.}$

$f_{pe} = 150,000 \text{ psi}$

$E_c = 4.3 \times 10^6 \text{ psi}$

Compute the depth of the neutral axis for the cracked section:

$$\frac{\lambda_1^u}{M} = \frac{1.07 \times 150,000 \times 8}{3 \times 7,445,560} = 0.057$$

$$b_w - \frac{\lambda_2}{M} = 8 - \frac{1.07 \times 150,000 \times 34.6}{7,445,560} = 2.033$$

$$\lambda_3^u = 2 \times (48 - 8) \times 5.75 + \frac{2 \times 1.8 \times 29 \times 10^6}{4.3 \times 10^6} + 0 = 484.279$$

$$\lambda_4^u = 1.07 \times 150,000 \Big[ 2 \times 40 \times 5.75 \times 34.6 - (40)(5.75)^2$$

$$- \frac{2 \times 1.8 \times 29 \times 10^6}{4.3 \times 10^6}(38 - 34.6) - 0 \Big]$$

$$\lambda_4^u = 2,329,007,661$$

$$\left( \lambda_3^u - \frac{\lambda_4^u}{M} \right) = 484.279 - \frac{2,329,007,661}{7,445,560} = 171.474$$

$$\lambda_5^u = 40 \times 5.75^2 + \frac{2 \times 1.8 \times 29 \times 38}{4.3^6} + 0$$

$$\lambda_5^u = 2245.10$$

$$\lambda_6^u = 1.07 \times 150,000 \times \Big[ \frac{2}{3}(48 - 8) \times 5.75^3 + \frac{2 \times 1.8 \times 29 \times 10^6}{4.3 \times 10^6}$$

$$(38 - 34.6) \times 38 + 0 - (48 - 8) \times 5.75^2 \times 34.6 \Big]$$

$$\lambda_6^u = -6,027,105,767$$

$$\left( \lambda_5^u - \frac{\lambda_6^u}{M} \right) = 1435.61$$

Therefore, equation for the depth of neutral axis, $kd$, is:

$$0.057(kd)^3 + 2.033(kd)^2 + 171.474(kd) - 1435.61 = 0$$

$$(kd) = 7.66\,\text{in.}$$

Maximum concrete stress:

$$f_{ct} = \frac{1.07 \times 150,000 \times 7.66}{\dfrac{48 \times 7.66^2}{2} - \dfrac{(40)}{2}(7.66 - 5.75)^2 - \dfrac{1.8 \times 29 \times 10^6}{4.3 \times 10^6}(38 - 7.66) - 0}$$

$$f_{ct} = 1274\,\text{psi}$$

$$f_s = \frac{1274 \times 29 \times 10^6}{4.3 \times 10^6}\left[\frac{38 - 7.66}{7.66}\right] = 34,082\,\text{psi}$$

Taking the moment about the center of gravity of the prestressed steel:

$$= \frac{8 \times 7.65 \times 1274}{2}\left(34.6 - \frac{7.65}{3}\right)$$

$$+ 40 \times 5.75 \times 1274 \left(\frac{7.65 - 5.75}{7.65}\right)\left(34.6 - \frac{5.75}{2}\right)$$

$$+ \frac{40 \times 5.75 \times 957.6}{2}\left(34.6 - \frac{5.75}{3}\right) + 1.8 \times 34,082(38 - 34.6)$$

$$= 7,400,0855\,\text{in.-lb}$$

$$\text{Error} = 1\%$$

Moment of inertia of cracked section:

$$I_{cr} = \frac{48 \times 7.66^3}{3} - \frac{(40)(7.66 - 5.75)^3}{3} + 1.8 \times \frac{29}{43}(38 - 7.66)^2$$

$$I_{cr} = 11,175\,\text{in.}^4$$

Note that the contribution of prestressed steel is not added to the section.

$$I_e = 11,175 + (82,065 - 11,175)\left(\frac{5,971,503}{7,445,560}\right)^3 = 47,757\text{in.}^4$$

$$\delta_{max} = \frac{5 \times 7,445,560 \times (70 \times 12)^2}{48 \times 4.3 \times 10^6 \times 47,757} - \frac{1.07 \times 150,000 \times (70 \times 12)^2}{8 \times 4.3 \times 10^6 \times 47,757} \times \frac{5}{6} \times 21.7$$

$$\delta_{max} = 2.66 - 1.25 = 1.41\,\text{in.}$$

*Table 6.2* Comparison of stresses for bonded and unbonded tendons, Examples 6.3 and 6.4

| Parameter | Tendon Configuration | |
|---|---|---|
| | Bonded | Unbonded |
| Depth of neutral axis | 9.15 in. | 7.66 in. |
| $f_{ct}$ | 1171 psi | 1274 psi |
| $f_s$ | 24,970 psi | 34,082 psi |
| $f_{ps}$ | 170,633 psi | 150,000 psi (assumed) |
| Deflection, $\delta$ | 1.10 in. | 1.41 in. |
| Cracking moment, $M_{cr}$ | 6,251,232 in.-lb | 5,971,503 in.-lb |
| Cracking moment, $M_n$ | 12,352,216 in.-lb | 7,432,181 in.-lb |

Estimate the increase in tendon stress due to deflection:

$$\frac{\delta}{e} = \frac{1.41}{21.7} = 0.065$$

$$\frac{e}{L} = \frac{21.7}{70 \times 12} = 0.026$$

$$\alpha_1 = 0.000923 \times 0.026 + 5.11 \times 0.026^2 = 0.0034$$

$$\alpha_2 = \left(11/10^7\right) e^{3.51} = 368/10^7$$

$$\Delta\varepsilon_{ps} = 0.0034 \times 0.065 + \frac{368}{10^7} \times (0.065)^2 = \frac{1}{10^6}(221 + 15.55)$$

$$\Delta f_{ps} = 27 \times 10^6 \times \frac{236.5}{10^6} = 6387 \text{ psi}$$

$$= 4.3\% \text{ of effective prestress}$$

If the calculations are repeated using $f_{ps} = 156,387$ psi instead of 150,000 psi, stresses in concrete and steel will decrease slightly. The comparison of stresses and deflections for bonded and unbonded tendons is shown in Table 6.2. Since the stress increase in unbonded tendons due to extra curvature is neglected, the stresses and deflections are higher.

## 6.10 Strength analysis: nominal moment capacity – unstrengthened beams

As in the case of reinforced concrete, the discussion presented in this section is a summary of provisions on allowable limits, basic principles, and computational procedures. More details can be found in books on prestressed concrete.

The basic principles used for the computation of the nominal strength of prestressed concrete beams are same as the ones used for reinforced concrete. The depth of the neutral axis at failure is determined using the force equilibrium and the moment capacity, and is computed using the moment equilibrium. The major difference is the presence of prestressed reinforcement and its stress level at failure. A typical T-section at failure is shown in Figure 6.9. At failure, the maximum strain in concrete is assumed to be 0.003 in./in. and the rectangular stress block is assumed valid. The contribution of compression steel should be computed by estimating the stress, $f'_s$, at failure. The stress will depend on the depth of the neutral axis and the location of the compression steel. In almost all cases, the nonprestressed tension reinforcement will yield and, hence, the stress for this steel can be assumed to be yield strength, $f_y$.

The stress level in prestressed reinforcement will depend on the effective prestress and curvature of the cross-section or overall deformed shape of the beam. Based on the strain level, the stresses can be estimated using approximate formulae or strain compatibility. The computation procedures are also different for bonded and unbonded tendons. The contribution of concrete:

$$C_c = \text{area of concrete in compression} \times 0.85 f'_c \tag{6.40}$$

For a flanged beam:

$$C = 0.85 f'_c b_w a + 0.85 f'_c \left(b - b_w\right) h_f \tag{6.41}$$

Contribution of compression steel:

$$C_s = A'_s f'_s = A'_s E'_s \left(\frac{\dfrac{a}{\beta_1} - d'}{\dfrac{a}{\beta_1}}\right) \leq A'_s f_y \tag{6.42}$$

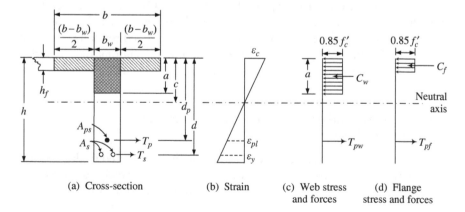

(a) Cross-section     (b) Strain     (c) Web stress and forces     (d) Flange stress and forces

*Figure 6.9* Strain, stress, and forces in flanged sections.

Contribution of nonprestressed tension steel:

$$T_s = A_s f_y \tag{6.43}$$

Contribution of prestressed steel:

$$T_p = A_{ps} f_{ps} \tag{6.44}$$

The value of $f_{ps}$ can be estimated using the following procedure, adopted from the ACI Code (2005).

### 6.10.1 Bonded tendons

If the effective prestress, $f_{pe}$, is greater than 50% of the fracture strength, $f_{pu}$, then:

$$f_{ps} = f_{pu} \left( 1 - \frac{\gamma_p}{\beta_1} \left\{ \rho_p \frac{f_{pu}}{f_c'} + \frac{d_s}{d_p} (\omega - \omega') \right\} \right) \tag{6.45}$$

where:

$$\rho_p = \frac{A_{ps}}{b d_p} \tag{6.46}$$

$$\omega = \frac{A_s \, f_y}{b d_s \, f_c'} \tag{6.47}$$

$$\omega' = \frac{A_s' \, f_y}{b d_s \, f_c'} \tag{6.48}$$

$d'$ should not be greater than $0.15 d_p$ and $\left[ \rho_p \frac{f_{pu}}{f_c'} + \frac{d_s}{d_p} (\omega - \omega') \right]$ should not be less than 0.17.

$\gamma_p = 0.55$ for $f_{py}/f_{pu}$ not less than 0.80 (high-strength prestressing bars)

$\gamma_p = 0.40$ for $f_{py}/f_{pu}$ not less than 0.85 (stress-relieved strands)

$\gamma_p = 0.28$ for $f_{py}/f_{pu}$ not less than 0.90 (low-relaxation strands)

### 6.10.2 Unbonded tendons

If the span-to-depth ratio, $L/d$, is less than or equal to 35:

$$f_{ps} = f_{pe} + 10{,}000 + \frac{f_c'}{100 \rho_p} \leq f_y \text{ or } \left( f_{pe} + 60{,}000 \right) \tag{6.49}$$

If $\dfrac{L}{d} > 35$:

$$f_{ps} = f_{pe} + 10,000 + \frac{f_c'}{300\rho_p} \leq f_{py} \text{ or } (f_{pe} + 30,000) \tag{6.50}$$

Once all the stresses are estimated, the depth of stress block, $a$, can be computed using:

$$C_c + C_s = T_s + T_p \tag{6.51}$$

$$0.85f_c'b_w a + 0.85f_c'(b - b_w) h_f + A_s'f_s' = A_s f_y + A_{ps}f_{ps} \tag{6.52}$$

The nominal moment capacity, $M_n$, can be computed by taking moment about the center of gravity of the prestressing steel.

$$M_n = 0.85f_c'b_w a \left(d_p - \frac{a}{2}\right) + 0.85f_c'(b - b_w) h_f \left(d_p - \frac{h_f}{2}\right)$$
$$+ A_s'f_s'(d_p - d') + A_s f_y (d_s - d_p) \tag{6.53}$$

### 6.10.3   Limitations on reinforcement

#### 6.10.3.1   Minimum reinforcement

If the prestressed reinforcement ratio, $\rho_p$, or $(\omega_p = \rho_p f_{ps}/f_c')$ is too small, when the section cracks, the reinforcement may not be able to carry the force that the tension zone concrete carried prior to cracking. This will result in abrupt and brittle failure. To prevent this type of failure, ACI Code (2005) restricts the moment capacity, $\phi M_n$, to the maximum of $1.2 M_{cr}$, or:

$$\phi M_n \geq 1.2M_{cr} \tag{6.54}$$

$M_{cr}$ is computed using a modulus of rupture of $7.5\sqrt{f_c'}$. If the flexural and shear capacity of the section exceeds twice the moments and shears calculated using factored loads, the restriction (Equation (6.50)) need not be followed. The minimum value for nonprestressed tension reinforcement, $A_s$, is $0.004\,A$. The $A$ is the part of the cross-section between the flexural tension face and the center of gravity of gross section. This reinforcement should be uniformly distributed over the precompressed tension zone and should be placed as close as possible to the extreme tension fiber. For two-way flat plates, if tensile stresses at service load exceed $2\sqrt{f_c'}$:

$$A_s \geq \frac{N_c}{0.5f_y} \tag{6.55}$$

where $N_c$ is the tensile force in the concrete due to unfactored dead plus live loads. The yield strength of steel, $f_y$, should be less than or equal to 60,000 psi.

For slabs at column support (tension on top):

$$A_s \geq 0.00075 \times hL \tag{6.56}$$

where $h$ is the slab thickness and $L$ is the span length parallel to the reinforcement being determined.

### 6.10.3.2  Maximum reinforcement

The maximum limit on reinforcement is placed to initiate the failure by yielding of steel, thus ensuring a ductile failure. Since a simple limit cannot be placed on prestressed steel area, the following restrictions are specified in ACI Code (2005).

For rectangular cross-section with prestress only:

$$\omega_p = \frac{A_{ps}\, f_{ps}}{b d_p\, f'_c} \leq 0.36 \beta_1 \tag{6.57}$$

If nonprestressed reinforcement is added:

$$\omega_p + \frac{d_s}{d_p}(\omega - \omega') = \left[ \frac{A_{ps}}{b d_p} \times \frac{f_{ps}}{f'_c} + \frac{d_s}{d_p}\left( \frac{A_s f_y}{b d_s f'_c} - \frac{A'_s f_y}{b d_s f'_c} \right) \right] \leq 0.36 \beta_1 \tag{6.58}$$

$$\frac{0.85a}{d_p} \leq 0.36 \beta_1 \tag{6.59}$$

For flanged sections, if $a$ is greater than flange thickness, $h_f$:

$$\omega_{pw} + \frac{d}{d_p}(\omega_w - \omega'_w) \text{ or } \frac{0.85a}{d_p} \leq 0.36 \beta_1 \tag{6.60}$$

If the external factored load moment does not exceed moment strength computed using only the compression portion of the moment couple, the maximum reinforcement limits need not be followed. The following inequalities will satisfy this condition.

For triangular sections:

$$\frac{M_u}{\phi} \leq f'_c b d_p^2 \left( 0.36 \beta_1 - 0.08 \beta_1^2 \right) \tag{6.61}$$

For flanged sections:

$$\frac{M_u}{\phi} \le f_c'bd_p^2 \left(0.36\beta_1 - 0.08\beta_1^2\right) + 0.85f_c' \left(b - b_w\right) h_f \left(d_p - 0.5h_f\right) \quad (6.62)$$

## 6.11   Computation of nominal moment capacity, $M_n$

### Example 6.5: Bonded tendons
For Example 6.3, compute the nominal moment capacity, $M_n$. Assume low-relaxation steel for prestressed tendons and the tendons are bonded.

Solution:

$$\begin{aligned}
b &= 48\,\text{in.}\\
b_w &= 8\,\text{in.}\\
h &= 40\,\text{in.}\\
d_s &= 38\,\text{in.}\\
d_p &= 34.6\,\text{in.}\\
A_s &= 1.8\,\text{in.}^2\\
A_{ps} &= 1.07\,\text{in.}^2\\
f_{pe} &= 150{,}000\,\text{psi}\\
f_{pu} &= 270{,}000\,\text{psi}\\
f_c' &= 5000\,\text{psi}\\
f_y &= 60{,}000\,\text{psi}
\end{aligned}$$

Computation of depth of stress block:
Contribution of concrete, assuming $c < h_f$:

$$C = 0.85f_c'ba$$

$$C = 0.85 \times 5000 \times 48a$$

$$C = 204{,}000a$$

Contribution of nonprestressed steel:

$$T_s = A_sf_y = 1.8 \times 60{,}000 = 108{,}000$$

$$T_s = 108{,}000\,\text{lb}$$

Contribution of prestressed steel:

$$T_p = A_{ps}f_{ps}$$

$$\gamma_p = 0.28, \quad \beta_1 = 0.8$$

$$f_{ps} = f_{pu}\left(1 - \frac{\gamma_p}{\beta_1}\left\{\rho_p\frac{f_{pu}}{f_c'} + \frac{d_s}{d_p}(\omega - \omega')\right\}\right)$$

$$f_{ps} = 270,000\left(1 - \frac{0.28}{0.8}\left\{\frac{1.07}{48 \times 34.6} \times \frac{270,000}{5000} + \frac{38}{34.6} \times \frac{1.8}{48 \times 38} \times \frac{60,000}{5000}\right\}\right)$$

$$f_{ps} = 265,483 \text{ psi}$$

$$T_p = 1.07 \times 265,483 = 284,067 \text{ lb}$$

$$C = T_s + T_p$$

Therefore, $204,000a = 108,000 + 284,067$

$$a = 1.92 < h_f = 5.75 \text{ in.}$$

$$M_n = 0.85f_c'ba\left(d_p - \frac{a}{2}\right) + A_sf_y(d_s - d_p)$$

$$M_n = 0.85 \times 5000 \times 48 \times 1.92\left(34.6 - \frac{1.92}{2}\right) + 1.8 \times 60,000(38 - 34.6)$$

$$= 13,543,315 \text{ in.-lb}$$

Check for minimum and maximum reinforcement:

$$M_{cr} = 6,251,232 \text{ in.-lb}$$

Assuming $\phi = 0.9$,

$$\phi M_n = 12,188,984 > 1.2M_{cr} = 7,166,464 \text{ in.-lb. Hence OK.}$$

$$0.004A = 0.004 \times 8 \times (40 - 12.9) = 0.8672 \text{ in.}^2$$

Note that $\bar{y} = 12.9$ in. (from Example 6.2):

$$A_s = 1.8 \text{ in.}^2 > 0.004A = 0.8672 \text{ in.}^2$$

$$\frac{0.85a}{d_p} = \frac{0.85 \times 6.15}{34.6} = 0.151 < 0.36\beta_1 = 0.288$$

Hence OK. Therefore nominal moment capacity, $M_n = 13,543,315$ in.-lb

### Example 6.6: Unbonded tendons
Repeat Example 6.5 assuming that tendons are unbonded.

Solution:

The difference is in the contribution of prestressed steel:

$$\frac{L}{d_p} = \frac{70' \times 12''}{34.6''} = 24.2 < 35,$$

Therefore, the stress in the unbonded tendon:

$$f_{ps} = f_{pe} + 10,000 + \frac{f'_c}{100\rho_p} \leq f_y \text{ or } \left(f_{pe} + 60,000\right)$$

$$\rho_p = \frac{A_{ps}}{bd_p} = \frac{1.07}{34.6 \times 48} = 0.00064$$

$$f_{ps} = f_{pe} + 10,000 + \frac{f'_c}{100\rho_p}$$

$$f_{ps} = 150,000 + 10,000 + \frac{5000}{100 \times 0.00064}$$

$$f_{ps} = 237,607 \text{ psi}$$

$$f_{pe} + 60,000 = 210,000 \text{ psi} < 237,607 < f_y$$

Therefore $f_{ps} = 210,000 \text{ psi}$

Computation of depth of stress block $a$:
Using the results of Example 6.5:

$$204,000a = 108,000 + 1.07 \times 210,000$$

Therefore $a = 1.63 \text{ in.} < h_f = 5.75 \text{ in.}$

Nominal moment capacity:

$$M_n = 0.55 \times 5000 \times 48 \times 1.63 \left(34.6 - \frac{1.63}{2}\right)$$

$$+ 1.8 \times 60,000 (38 - 34.6)$$

$$M_n = 11,601,388 \text{ in.-lb}$$

$$\phi M_n = 10,441,249 > 1.2M_{cr} = 7,166,464 \text{ in.-lb}$$

Since the reinforcement satisfies the minimum and maximum reinforcement ratios, nominal moment capacity:

$$M_n = 11,601,388 \text{ in.-lb}$$

## 6.12  Load–deflection behavior of typical strengthened prestressed concrete beams

The load–deflection behavior of a typical strengthened beam, shown in Figure 6.10, is similar to the unstrengthened beams with the following differences. The stiffness of the cracked section will increase due to the contribution of the composite reinforcement. The composite can be considered as

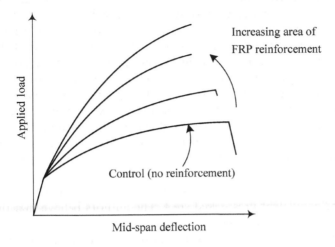

*Figure 6.10* Load–deflection behavior of a typical strengthened beam.

additional nonprestressed reinforcement. The strength capacity of the section will also be higher.

The behavior of uncracked section will not change significantly due to the addition of the composite, even if the composite was applied with no loads present. Cracked sections can be analyzed using procedures developed for unstrengthened sections.

The composite could affect the behavior of the beam significantly at failure. In most cases, the failure of prestressed concrete beams, initiated by yielding of reinforcement, occur by crushing of concrete. The addition of composite could significantly alter the initiation of failure, either by fracture or by debonding of composite. The various mechanisms are discussed in Section 6.15. Limited experimental results indicate that the addition of composite does not decrease the failure deflection significantly (Cha *et al.*, 1999). Therefore, it could be assumed that strengthening with composites does not affect the ductile behavior of prestressed concrete beams.

## 6.13 Analysis of cracked sections: strengthened beams

As mentioned earlier, the analysis of strengthened sections is similar to the analysis of unstrengthened sections presented in Section 6.8. The contribution of composite is similar to the contribution of nonprestressed reinforcement. It is assumed that composite will be used to increase tensile force contribution. If there is a need, the composite can also be used to increase the compression force capacity. However, it should be verified that the thin composite plate would not delaminate and buckle. The cracked section analysis can be summarized in the following steps.

*Step 1*: Compute the stress and strain in nonprestressed tension steel and strain at the location where composite will be attached. These strains should be computed for loads present when renovation will be carried out. Since the stresses and strains are moment dependent, only if the strain at composite location is zero, during the application of composite, equations of Sections 6.13.1 and 6.13.2 will provide perfect solutions. Since most of the live load will be removed during renovation, the stress (strain) will be near zero at the extreme tension face where the composite will be attached. However, solutions should be considered approximate. The magnitude of the error will depend on the magnitude of the strain at the composite location during renovation. For almost all cases, the solution can be considered as accurate for design purposes.

If a perfect solution is needed, the area of composite can be converted to the equivalent prestressed steel. The authors believe that the complexity of such a solution is not warranted.

*Step 2*: Compute the stresses and deflections using the modified equations incorporating the composite reinforcement (Sections 6.13.1 and 6.13.2).

### 6.13.1　*Equations for strengthened sections: bonded tendons*

The equation for computing depth of neutral axis is the same as Equation (6.11) but the coefficients ($\lambda_s$) are different. Using additional subscripts "$c$" for composites, the equation for computing $kd$ is:

$$\frac{\lambda_{1c}}{M}(kd)^3 + \left(b_w - \frac{\lambda_{2c}}{M}\right)(kd)^2 + \left(\lambda_{3c} - \frac{\lambda_{4c}}{M}\right)(kd) - \left(\lambda_{5c} + \frac{\lambda_{6c}}{M}\right) = 0$$

(6.63)

$$\lambda_{1c} = \lambda_1$$

(6.64)

For $\lambda_1$ to $\lambda_6$, use Equations (6.12)–(6.17):

$$\lambda_{2c} = \lambda_2$$

(6.65)

$$\lambda_{3c} = \lambda_3 + 2\frac{A_f E_f}{E_c}$$

(6.66)

where $A_f$ is the area of composite and $E_f$ is the modulus of the composite. As in the case of reinforced concrete, it is recommended to use equivalent fiber area and fiber modulus for $A_f$ and $E_f$.

$$\lambda_{4c} = \lambda_4 - 2A_{ps}E_{ps}\left(\varepsilon_{pe} + \varepsilon_{ce}\right)\frac{A_f E_f}{E_c}\left(d_f - d_p\right)$$

(6.67)

where $d_f$ is the depth measured to the center of gravity of the composite reinforcement.

$$\lambda_{5c} = \lambda_5 + \frac{2A_f E_f d_f}{E_c} \tag{6.68}$$

$$\lambda_{6c} = \lambda_6 + 2A_{ps}E_{ps}\left(\varepsilon_{pe} + \varepsilon_{ce}\right)\frac{A_f E_f}{E_c}\left(d_f - d_p\right) \tag{6.69}$$

$$f_{ct} = \cfrac{A_{ps}E_{ps}(\varepsilon_{pe} + \varepsilon_{ce})(kd)}{\left[\begin{array}{c}\dfrac{b(kd)^2}{2} - \dfrac{(b - b_w)(c - h_f)^2}{2} - \dfrac{A_{ps}E_{ps}}{E_c}(d_p - kd) - \\[3mm] \dfrac{A_s E_s}{E_c}(d_s - kd)\dfrac{A_s' E_s'}{E_c}(d_s' - kd) - \dfrac{A_f E_f}{E_c}(d_f - kd)\end{array}\right]} \tag{6.70}$$

Estimate maximum stresses in nonprestressed steel and prestressed steel using Equations (6.19)–(6.21). The stress in composite:

$$f_f = \frac{E_f}{E_c}f_{ct}\left(\frac{d_f - kd}{kd}\right) \tag{6.71}$$

Moment of inertia:

$$I_{cr} = (I_{cr} \text{ from Equation (6.22))} + A_f \frac{E_f}{E_c}(d_f - kd)^2 \tag{6.72}$$

Use this $I_{cr}$ and Equation (6.23) to compute the new effective moment of inertia.

### 6.13.2  Equations for strengthened section: unbonded tendons

Using the same logic as used for the unbonded tendon:

$$\lambda_{1c}'' = \lambda_1'' \tag{6.73}$$

$$\lambda_{2c}'' = \lambda_2'' \tag{6.74}$$

$$\lambda_{3c}'' = \lambda_3'' + 2\frac{A_f E_f}{E_c} \tag{6.75}$$

$$\lambda_{4c}'' = \lambda_4'' - 2A_{ps}f_{pe}\frac{A_f E_f}{E_c}(d_f - d_p) \tag{6.76}$$

$$\lambda_{5c}'' = \lambda_5'' + \frac{2A_f E_f d_f}{E_c} \tag{6.77}$$

$$\lambda_{6c}'' = \lambda_6'' + 2A_{ps}f_{pe}\frac{A_f E_f}{E_c}(d_f - d_p) \tag{6.78}$$

Use these $\lambda$ variables and Equation (6.28) for computing the depth of the neutral axis. The maximum concrete stress:

$$f_{ct} = \frac{A_{ps}f_{pe}(kd)}{\left[\begin{array}{c} \dfrac{b(kd)^2}{2} - \dfrac{(b-b_w)(kd-h_f)^2}{2} - \dfrac{A_sE_s}{E_c}(d_s-kd)- \\[2mm] \dfrac{A'_sE'_s}{E_c}(d'_s-kd) - \dfrac{A_fE_f}{E_c}(d_f-kd) \end{array}\right]} \tag{6.79}$$

Stresses in nonprestressed steel can be computed using Equations (6.37) and (6.38).

For the moment of inertia of the cracked section, add the contribution of the composite.

$$I_{cr} = (I_{cr} \text{ from Equation (6.39)}) + A_f\frac{E_f}{E_c}(d_f-kd)^2 \tag{6.80}$$

Equation for effective moment of inertia, $I_e$ is the same as Equation (6.23).

## 6.14   Computation of stresses and deflection for strengthened beams

### *Example 6.7: Bonded tendons*
For the T-beam of Example 6.3, compute maximum stresses and deflection. When the beam was strengthened, only the dead load moment of 4,505,560 in.-lb was present. The section was strengthened using three layers of 8-in. wide carbon fibers. Equivalent fiber thickness = 0.0065 in. and fiber modulus = $33 \times 10^6$ psi. The maximum moment = 9,000,000 in.-lb.

Solution:

Using the results of Examples 6.2 and 6.3, for the moment of 4,505,560 in.-lb, the section does not crack.

$$A_g = 550 \text{ in.}^2$$

$$I_g = 82,065 \text{ in.}^4$$

$$\bar{y} = y_t = 12.9 \text{ in.}$$

$$f_{ps} = f_{pe} \text{ (since moment} = M_D)$$

$$f_{ct} = \frac{M}{I_g}y_t + \frac{A_{ps}f_{sc}}{A_g} - \frac{A_{ps}+\delta_e e_0}{I_g}y_t$$

$$f_{ct} = \left(\frac{4,505,560 - 1.07\times150,000\times21.7}{82,065}\right)\times12.9 + \frac{1.07\times150,000}{550}$$

$f_{ct} = 453\,\text{psi}$

$$f_s = \frac{29}{4.3}\left[-\left(\frac{4,505,560 - 1.07 \times 150,000 \times 21.7}{82,065}\right) \times (38 - 12.9)\right.$$
$$\left. + \left(\frac{1.07 \times 150,000}{550}\right)\right]$$

$f_s = 142\,\text{psi (tension)}$

$\varepsilon_s = 4.9 \times 10^{-6}\,\text{in./in.} \cong 0$

$$f_b = \left[-\left(\frac{4,505,560 - 1.07 \times 150,000 \times 21.7}{82,065}\right) \times (40 - 12.9)\right.$$
$$\left. + \frac{1.07 \times 150,000}{550}\right]$$

$f_b = 46\,\text{psi (tension)}$

$$\varepsilon_{bi} = \frac{46}{4.3 \times 10^6} = 1.1 \times 10^{-5}\,\text{in./in.} \cong 0$$

Therefore the solution can be considered accurate.
Deflection:

$$\delta_{max} = \left(\frac{5 \times 4,505,560 \times (70 \times 12)^2}{48 \times 4.3 \times 10^6 \times 82,065}\right) - \left(\frac{1.07 \times 150,000 \times (70 \times 12)^2}{8 \times 4.3 \times 10^6 \times 82,065}\right)$$
$$\times \frac{5}{6} \times 21.7$$

$$= 0.94 - 0.73 = 0.21$$

$\delta_{max} = 0.21\,\text{in.} \downarrow \text{(downward)}$

When carbon composite was applied, external moment $= 4,505,560\,\text{in.-lb}$

$f_{ps} = f_{se} = 150,000\,\text{psi}$

$f_s = 142\,\text{psi (tension)}$

$f_{ct} = 453\,\text{psi (compression)}$

$\varepsilon_{bi} \cong 0$

$\delta = 0.21\,\text{in.}$

Strengthened section:
Area of composite, $A_f = 3 \times 8 \times 0.0065 = 0.156\,\text{in.}^2$

$E_f = 33 \times 10^6\,\text{psi}$

Compute the depth of the neutral axis of the cracked (strengthened) section.

Using the results of Example 6.3:

$$\lambda_{1c} = 428{,}342$$

$$\lambda_{2c} = 44{,}785{,}043$$

$$\lambda_{3c} = 497.8 + \frac{2 \times 0.156 \times 33 \times 10^6}{4.3 \times 10^6} = 500.2$$

$$\lambda_{4c} = 2{,}344{,}361{,}141 - 2 \times 1.07 \times 27 \times 10^6 \times 0.00556$$

$$\times \frac{0.156 \times 33 \times 10^6}{4.3 \times 10^6} \times (40 - 34.6)$$

$$\lambda_{4c} = 2{,}342{,}284{,}238$$

$$\lambda_{5c} = 2710 + \frac{2 \times 0.156 \times 33 \times 10^6 \times 40}{4.3 \times 10^6} = 2806$$

$$\lambda_{6c} = 6{,}031{,}520{,}292 + 2 \times 1.07 \times 10^6$$

$$\times 0.00556 \times \frac{0.156 \times 33 \times 10^6}{4.3 \times 10^6} \times (40 - 34.6)$$

$$= -6{,}029{,}443{,}389$$

Therefore the cubic equation for $kd$ is:

$$0.0472\,(kd)^3 + 3.02\,(kd)^2 + 240(kd) - 2136 = 0$$

$$kd = 8.0\,\text{in.} > h_f = 5.75\,\text{in. Hence OK.}$$

Moment of inertia of cracked section:

$$I_{cr} = \left(\frac{48 \times 8^3}{3}\right) - \left(\frac{40}{3}\right)(8 - 5.75)^3 + 1.07 \times \left(\frac{27}{4.3}\right)(34.6 - 8)^2$$

$$+ 1.8 \times \left(\frac{29}{4.3}\right)(38 - 8)^2 + 0.156 \times \left(\frac{33}{4.3}\right)(40 - 8)^2 \text{ in.}^4$$

Using Equation (6.70), maximum concrete stress:

$$f_{ct} = \frac{1.07 \times 27 \times 10^6 \times 0.00556 \times 8}{\dfrac{48 \times 8^2}{2} - \dfrac{40(8 - 5.75)^2}{2} - \dfrac{1.07 \times 27(34.6 - 8)}{4.3} - \dfrac{1.8 \times 29(38 - 8)}{4.3} - \dfrac{0.156 \times 33(40 - 8)}{4.3}}$$

$$f_{ct} = 1505\,\text{psi}$$

Using Equation (6.19):

$$f_s = \left(\frac{29}{4.3}\right) \times 1505 \left(\frac{38-8}{8}\right) = 38,076 \, \text{psi}$$

Using Equation (6.21):

$$f_{ps} = 27 \times 10^6 (0.00556) + \left(\frac{27}{4.3}\right) \times 1505 \left(\frac{34.6-8}{8}\right)$$

$$f_{ps} = 181,541 \, \text{psi}$$

Using Equation (6.71):

$$f_f = \left(\frac{33}{4.3}\right) \times 1505 \left(\frac{40-8}{8}\right) = 46,200 \, \text{psi}$$

$$M_{cr} = 6,251,232 \, \text{in.-lb}$$

Therefore:

$$I_e = 25,687 + (82,065 - 25,687)\left(\frac{6,251,232}{9,000,000}\right)^3 = 44,579 \, \text{in.}^4$$

Deflection:

$$\delta = \left(\frac{5 \times 9,000,000 \times (70 \times 12)^2}{48 \times 4.3 \times 10^6 \times 44,579}\right) - \left(\frac{1.07 \times 181,541 \times (70 \times 12)^2}{8 \times 4.3 \times 10^6 \times 44,579}\right)$$

$$\times \frac{5}{6} \times 21.7$$

$$= 3.45 - 1.62 = 1.83$$

$$\delta = 1.83 \, \text{in.}$$

### Example 6.8: Unbonded tendons

Repeat Example 6.7 assuming unbonded tendons.

Solution:

Since $f_{ps} = f_{pe}$ for Example 6.7, the bottom strain,

$$f_{bi} \cong 0$$

Computation of $kd$:

Using Equations (6.73)–(6.78) and Example 6.4:

$$\lambda''_{1c} = 428,000, \frac{\lambda''_{1c}}{M} = 0.0476$$

$$\lambda''_{2c} = 44,426,400, \left(b_w - \frac{\lambda''_{2c}}{M}\right) = 3.064$$

$$\lambda''_{3c} = 484.279 + \frac{2 \times 0.156 \times 33 \times 10^6}{4.3 \times 10^6} = 486.67$$

$$\lambda''_{4c} = 2,329,007,661 - \frac{2 \times 1.07 \times 150,000 \times 0.156 \times 33(40 - 34)}{4.3}$$

$$\lambda''_{4c} = 2,326,932,418$$

$$\left(\lambda''_{3c} - \frac{\lambda''_{4c}}{M}\right) = 228.12$$

$$\lambda''_{5c} = 2245.10 + \frac{2 \times 0.156 \times 33 \times 10^6 \times 40}{4.3 \times 10^6} = 2340.90$$

$$\lambda''_{6c} = -6,027,105,767 + \left(\frac{2 \times 1.07 \times 150,000 \times 0.156 \times 33 \times 10^6}{4.3 \times 10^6}\right)$$

$$\times (40 - 34)$$

$$\lambda''_{6c} = -6,025,030,524$$

$$\left(\lambda''_{5c} - \frac{\lambda''_{6c}}{M}\right) = 1671.5$$

Therefore $0.0476(kd)^3 + 3.064(kd)^2 + 228.12(kd) - 1671.5 = 0$

$kd = 6.7\,\text{in.}$

$$f_{ct} = \frac{1.07 \times 150,000 \times 6.7}{\frac{48 \times 6.7^2}{2} - \frac{40(6.7 - 5.75)^2}{2} - \frac{1.8 \times 29(38 - 6.7)}{4.3} - \frac{0.156 \times 33(40 - 6.7)}{4.3}}$$

$f_{ct} = 1682\,\text{psi}$

$$f_s = \frac{29}{4.3} \times 1682 \left(\frac{38 - 6.7}{6.7}\right) = 52,994\,\text{psi}$$

$$f_f = \frac{33}{4.3} \times 1682 \left(\frac{40 - 6.7}{6.7}\right) = 64,157\,\text{psi}$$

$$I_{cr} = \frac{48 \times 6.7^3}{3} - \frac{40}{3}(6.7 - 5.75)^3 + 1.8 \times \frac{29}{4.3}(38 - 6.7)^2 + 0.156$$

$$\times \frac{33}{4.3}(40 - 6.7)^2$$

$I_{cr} = 18,021\,\text{in.}^4$

Therefore $I_e = 18,021 + (82,065 - 18,021)\left(\dfrac{5,971,503}{9,000,000}\right)^3$

$I_e = 36,728 \text{ in.}^4 \quad \delta_{max} = \dfrac{5 \times 9,000,000 \times (70 \times 12)^2}{48 \times 4.3 \times 10^6 \times 36,728}$

$$- \dfrac{1.07 \times 150,000 \times (70 \times 12)^2}{8 \times 4.3 \times 10^6 \times 36,728} \times \left(\dfrac{5}{6} \times 21.7\right)$$

$\delta_{max} = 4.18 - 1.63 = 2.55$

$\delta_{max} = 2.55 \text{ in. } \downarrow \text{ (downward)}$

If necessary, the increase in prestress can be calculated for this deflection and the calculations can be repeated using the new $f_{ps}$ (instead of $f_{pe}$).

## 6.15 Strength analysis: nominal moment capacity, $M_n$ – strengthened beams

For the most generalized case, strengthened beams have a compressive force contribution from concrete and compression steel, and a tension force contribution from nonprestressed steel, prestressed steel, and the composite. Linear strain distribution across the thickness can still be assumed valid. Typical strain, stress, and force contributions are shown in Figure 6.11.

For strengthened beams, it is most probable that the failure in the tension zone will be initiated by failure of composite because prestressed reinforcement and nonprestressed reinforcement can withstand much larger strains before failure. Therefore, the following sequence is recommended for computations. The basic principles are same; the depth of the neutral axis is computed using force equilibrium whereas the moment capacity is computed using moment equilibrium. For computing the force equilibrium, the strain

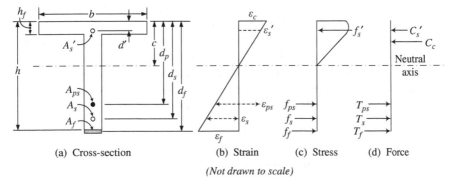

| (a) Cross-section | (b) Strain | (c) Stress | (d) Force |

*(Not drawn to scale)*

*Figure 6.11* Stress and strain distribution for a typical strengthened section at ultimate load.

and stress in concrete and reinforcement have to be computed using the strains at failure. The possible failure mechanisms are:

(a)  Failure of composite followed by crushing of concrete
(b)  Crushing of concrete followed by failure of composite

Failure of composite can occur by fracture or delamination, depending on the thickness. If failure is initiated by failure of composite, the maximum strain in concrete could be less than 0.003 in./in. and, hence, rectangular stress block assumption may not be valid. A nonlinear analysis is needed to compute the force contribution of concrete. The force contribution of prestressed steel must also be computed using nonlinear analysis. The nonprestressed tension reinforcement can be expected to yield and, hence, the stress in tension steel can be assumed to be yield strength. The stress in compression steel must be computed using strain compatibility.

The following procedure covers both failure mechanisms in a systematic way.

*Step 1*: Compute the maximum permissible strain in the composite, based on the thickness of the composite. As in the case of reinforced concrete, the fracture strains guaranteed by the manufacturer, $\varepsilon_{fu}$, could not be used for computations because failure might occur due to delamination. Using the principles adopted for reinforced concrete, maximum effective fracture strain of composite at failure,

$$\varepsilon_{fe} = \kappa_m \varepsilon_{fu} \leq 0.9 \tag{6.81}$$

$$\kappa_m = 1 - \frac{nE_f t_f}{2,400,000} \text{ for } nE_f t_f \leq 1,200,000 \tag{6.82}$$

$$\kappa_m = \frac{600,000}{nE_f t_f} \text{ for } nE_f t_f > 1,200,000 \tag{6.83}$$

As in the case of reinforced concrete, it is recommended to use equivalent fiber thickness and modulus for $t_f$ and $E_f$, respectively; $n$ is the number of layers. If prefabricated plates are used, $nt_f$ should be replaced with the equivalent fiber thickness of the entire plate.

*Step 2*: Assuming failure occurs simultaneously in concrete and composite, estimate strains, stresses, and forces. Assuming a failure strain of 0.003 in./in. for concrete, force contribution of concrete,

$$C_c = 0.85 f_c' b_w \beta_1 c + 0.85 f_c' (b - b_w) h_f \tag{6.84}$$

Contribution of compression steel:

$$C_s' = A_s' E_s' \times 0.003 \left( \frac{c - d'}{c} \right) \leq A_s' f_y \tag{6.85}$$

Contribution of nonprestressed tension steel:

$$T_s = A_s f_y \tag{6.86}$$

Contribution of composite:

$$T_f = A_f E_f \varepsilon_{fe} \tag{6.87}$$

The contribution of prestressed steel is estimated using the nonlinear relationship between stress and strain, recommended by Prestressed Concrete Institute (1999).

Strain in prestressed steel at failure:

$$\varepsilon_{ps} = \varepsilon_{pe} + \varepsilon_{ce} + 0.003 \left( \frac{d_p - c}{c} \right) \tag{6.88}$$

For grade 270 tendons:

$$f_{ps} = E_{ps} \varepsilon_{ps} \text{ for } \varepsilon_{ps} \leq 0.008 \tag{6.89a}$$

$$f_{ps} = f_{pu} - \left( \frac{75}{\varepsilon_{ps} - 0.0065} - 2000 \right) \text{ for } \varepsilon_{ps} > 0.008 \tag{6.89b}$$

For grade 250 tendons:

$$f_{ps} = E_{ps} \varepsilon_{ps} \text{ for } \varepsilon_{ps} \leq 0.008 \tag{6.90a}$$

$$f_{ps} = f_{pu} - \left( \frac{58}{\varepsilon_{ps} - 0.006} - 2000 \right) \text{ for } \varepsilon_{ps} > 0.008 \tag{6.90b}$$

$$T_p = A_{ps} f_{ps} \tag{6.91}$$

For force equilibrium:

$$C = C_c + C'_s = T = T_s + T_f + T_p \tag{6.92}$$

Equation (6.92) is valid for simultaneous failure of composite and crushing of concrete.

If $C > T$, then failure occurs by failure of composite. Then go to step 3.

If $C < T$, then failure occurs by crushing of concrete. Then go to step 4.

*Step 3*: Compute the depth of the neutral axis for the condition $C > T$. Maximum strain in composite, $\varepsilon_{fe}$.

$$\text{The strain at the composite level} = (\varepsilon_{fe} + \varepsilon_{bi}) \tag{6.93}$$

where $\varepsilon_{bi}$ is the strain at the level of the composite during renovation. $\varepsilon_{bi}$ should be computed using the properties of cracked or uncracked section, presented in the previous sections. Note that $\varepsilon_{bi}$ could be positive or negative depending on whether the strain at the location of the composite was tension or compression. Compressive strain should be taken as negative.

Using the results of balanced failure, assume a maximum strain in concrete. This maximum strain for concrete, $\varepsilon_c$, should be less than 0.003 in./in.

Using $\varepsilon_c$, $(\varepsilon_{fe} + \varepsilon_{bi})$, and similar-triangles principles, compute the strains, stresses, and forces in nonprestressed and prestressed steel and check for force equilibrium.

Using $T_f = A_f E_f \varepsilon_{fe}$, the same as Equation (6.87) using similar triangles, the depth of the neutral axis, $c$ (Figure 6.11):

$$c = \left(\frac{\varepsilon_c}{\varepsilon_c + \varepsilon_{fe}}\right) d_f \tag{6.94}$$

where $d_f$ is the distance between extreme compression fiber and composite.

$$C'_s = A'_s f'_s = A'_s E_s \varepsilon_c \left(\frac{c - d'}{c}\right) \leq A'_s f_y \tag{6.95}$$

$$T_s = A_s f_s = A_s E_s \varepsilon_c \left(\frac{d_s - c}{c}\right) \leq A_s f_y \tag{6.96}$$

$$\varepsilon_{ps} = \varepsilon_{pe} + \varepsilon_{ce} + \varepsilon_c \left(\frac{d_p - c}{c}\right) \tag{6.97}$$

$$T_p = A_{ps} \times Stress \tag{6.98}$$

Note that stress is a function of $\varepsilon_{ps}$, Equation (6.89) or (6.90).
Contribution of concrete:

$$C_c = \gamma f'_c \beta_1 \text{ (compression up to a height of } c) \tag{6.99}$$

$$\gamma = \frac{0.9 \ln\left[1 + \left(\frac{\varepsilon_c}{\varepsilon'_c}\right)^2\right]}{\beta_1 \left(\frac{\varepsilon_c}{\varepsilon'_c}\right)} \tag{6.100}$$

$$\beta_1 = 2 - \frac{4\left[\left(\frac{\varepsilon_c}{\varepsilon'_c}\right) - \tan^{-1}\left(\frac{\varepsilon_c}{\varepsilon'_c}\right)\right]}{\left(\frac{\varepsilon_c}{\varepsilon'_c}\right) \ln\left[1 + \left(\frac{\varepsilon_c}{\varepsilon'_c}\right)^2\right]} \tag{6.101}$$

where $\varepsilon'_c = \dfrac{1.71f'_c}{E_c}$

If $C_c + C'_s > T_s + T_f + T_p$, then the assumed maximum concrete strain is larger than the actual strain. Assume a smaller strain. If the total tension force is smaller than the total compression force, assume a larger strain. Iterate until equilibrium (or $C = T$) is satisfied.

*Step 4*: Compute the depth of the neutral axis for the condition $C < T$.

For this case, the maximum strain in concrete is 0.003 in./in., but the strain in the composite is less than $\varepsilon_{fe}$.

Assume a value for the depth of the neutral axis, $c$, which is greater than $\left(\dfrac{0.003}{\varepsilon_{fe} + 0.003}\right) d_f$, compute the forces, and check for equilibrium. Iterate until equilibrium is reached.

$C_c = 0.85f'_c \times$ (compression area up to a height of $\beta_1 c$).

Equations for $C_s$, $T_s$, and $T_p$ are the same as in step 3. The maximum strain in concrete, $\varepsilon_c$, is 0.003.

$$T_f = A_f E_f 0.003 \left(\frac{d_f - c}{c}\right) \tag{6.102}$$

*Step 5*: Once the depth of neutral axis is established, moment capacity can be computed using moment equilibrium. Again, as in the case of reinforced concrete, the contribution of the composite is multiplied by a factor $\psi$ for improved safety. Based on current knowledge, a factor of 0.85 is recommended for $\psi$. For consistency, the moments should be taken about the neutral axis.

Therefore $M_n = \gamma f'_c \beta_1 b_w c \left(c - \frac{\beta_1 c}{2}\right) + \gamma f'_c \beta_1 (b - b_w) b_f \left(c - \frac{b_f}{2}\right)$
$+ A'_s f'_s (c - d') + A_s f_s (d_s - c) + A_{ps} f_{ps} (d_p - c) \tag{6.103}$
$+ \psi A_f f_f (d_f - c)$

It is assumed that $a > b_f$. Otherwise, treat the beam as a rectangular section, so that $b$ and $b_w$ are the same.

*Step 6*: Check for minimum and maximum reinforcement. Using the ACI guidelines for prestressed concrete (2005), $\phi M_n$ should be greater than $1.2M_{cr}$, and $(A_s + A_f)$ should be greater than $0.004A$ to satisfy the minimum nonprestressed area requirements. For maximum reinforcement, Equation (6.59) or (6.60) can be used with the recomputed (with composite) depth of stress block, $a$, and factor $\beta_1$.

### 6.15.1 Special considerations for unbonded tendons

The procedure outlined for bonded tendons can be used for unbonded tendons with a modification for the computation of prestress, $f_{ps}$. As discussed earlier, for beams with unbonded tendons, the prestress at failure is beam dependent rather than section dependent. The following procedure, based on the reports of ACI Committee 423 on prestressed concrete (2005), is recommended for the prestress, $f_{ps}$. The equations cover both single and multiple spans.

For approximate analysis and initial design, assume:

$$f_{ps} = f_{pe} + 15,000 \text{ psi} \qquad (6.104)$$

For a more accurate analysis, compute the strain at the prestressed steel level and estimate the stress based on beam geometry.

$$f_{ps} = f_{pe} + \Omega_u E_{ps} \varepsilon_c \left( \frac{d_p}{c} - 1 \right) \left( \frac{L_1}{L_2} \right) \le 0.8 f_{pu} \qquad (6.105)$$

$\varepsilon_c$ is the maximum strain in concrete. If the failure is initiated by crushing of concrete, $\varepsilon_c$ is 0.003. If the failure is initiated by fracture of composite, $\varepsilon_c$ is less than 0.003.

$L_1$ is the length of the loaded span or spans affected by the tendon.

$L_2$ is the total length of the tendon between anchorages.

Note that Equation (6.105) applies to tendons used in multiple span beams and slabs. If the beam is a single span beam, then $L_1$, $L_2$, and the span length, $L$, are the same. The coefficient, $\Omega_u$, which depends on the beam span-to-depth ratio and type of loading, can be estimated using the following equations.

For uniform and third-point loading:

$$\Omega_u = 3 \left( \frac{d_p}{L} \right) \qquad (6.106)$$

For center-point loading:

$$\Omega_u = 1.5 \left( \frac{d_p}{L} \right) \qquad (6.107)$$

Once $f_{ps}$ is known, the contribution of prestressed tendons can be estimated for the computation of the neutral axis and moment capacity (Equations (6.98) and (6.103), respectively).

## 6.16  Computation of nominal moment, $M_n$, strengthened sections

### Example 6.9: Bonded tendons
Compute the nominal moment capacity for section of Example 6.7. Use a guaranteed fracture strain of 0.0167 in./in. for the composite.

Solution:

Using the results of Example 6.7:

$b = 48$ in.

$b_w = 8$ in.

$d_f = h = 40$ in.

$d_s = 38$ in.

$d_p = 34.6$ in.

$A_s = 1.2$ in.$^2$

$A_{ps} = 1.07$ in.$^2$

$A'_s = 0$ in.$^2$

$A_f = 0.156$ in.$^2$

When the composite was applied, strains $\varepsilon_{bi} = 0$, $\varepsilon_{ce} = 0.0036$ in./in.

$f_{pe} = 150,000$ psi

$f_{pu} = 270,000$ psi

$f'_c = 5000$ psi

$f_y = 60,000$ psi

$nE_f t_f = 3 \times 33 \times 10^6 \times 0.0065 = 643,500$

Using Equation (6.82):

$$\kappa_m = 1 - \frac{3 \times 33 \times 10^6 \times 0.0065}{2,400,000} = 0.732 < 0.9$$

Therefore effective failure strain of composite is computed using Equation (6.81):

$\varepsilon_{fe} = 0.732 \times 0.0167 = 0.0122$ in./in.

For balanced failure (simultaneous failure of concrete and composite), refer to Figure 6.12.

$\varepsilon_{cu} = 0.003$ in./in.

$\varepsilon_{fe} = 0.0122$ in./in.

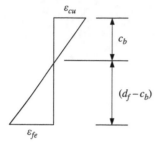

*Figure 6.12* Assumed balanced strain condition.

Since $\varepsilon_{bi} = 0$, $\dfrac{\varepsilon_{cu}}{\varepsilon_{fe}} = \dfrac{c_b}{(d_f - c_b)}$ or $\dfrac{c}{40 - c} = \dfrac{0.003}{0.0122}$

$c_b = 7.9\,\text{in.}$

$\beta_1 = 0.8$

Using Equation (6.84).

$C_c = 0.85 \times 5000 \times 8 \times 0.8 \times 7.9 + 0.85 \times 5000(40) \times 5.75$

$C_c = 1{,}192{,}380\,\text{lb}$

$C_s' = 0\,\text{lb}$

$T_s = 1.8 \times 60{,}000 = 108{,}000\,\text{lb}$

$T_f = 0.156 \times 33 \times 10^6 \times 0.0122 = 62{,}806\,\text{lb}$

Using Equation (6.88):

$\varepsilon_{ps} = \dfrac{150{,}000}{27 \times 10^6} + 0.00036 + 0.003\left(\dfrac{39.6 - 7.9}{7.9}\right)$

$\varepsilon_{ps} = 0.0161 > 0.008\,\text{in./in.}$

Using Equation (6.89b):

$f_{ps} = 270{,}000 - \dfrac{75}{0.0161 - 0.0065} - 2000$

$f_{ps} = 260{,}188\,\text{psi}$

$T_p = 1.07 \times 260{,}188 = 278{,}401\,\text{lb}$

Therefore $C = 1,192,380 \, lb$

$T = 108,000 + 62,806 + 278,401 = 449,207 \, lb$

$C > T$

Therefore failure is initiated by failure of composite.

$\varepsilon_{cu} < 0.003 \, in./in.$

$\varepsilon_{fe} = 0.0122 \, in./in.$

If $C_c = 449,207 \, lb = T$, the depth of the stress block will be less than $h_f$. If the maximum stress is $0.85 f_c'$:

$$a = \frac{449,207}{0.85 \times 5000 \times 48} = 2.2 \, in.$$

$$c = \frac{2.2}{0.8} \times 2.75 \, in.$$

Using Figure 6.13, the maximum compressive strain can be computed as:

$$\varepsilon_c = \frac{0.0122 \times 2.75}{(40 - 2.75)} = 0.0009 \, in./in.$$

Assume a maximum strain of $0.0011 \, in./in$ (Figure 6.14),

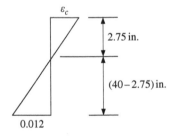

0.012

*Figure 6.13* Strain distribution across cross-section for Example 6.9.

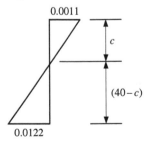

0.0122

*Figure 6.14* Strain distribution assuming maximum strain of 0.0011.

$$\frac{c}{0.0011} = \frac{40 - c}{0.122}$$

or $c = 3.3$ in. $< h_f = 5.75$ in.

$\beta_1 = 0.7$ (Equation (6.101))

$a = 0.7 \times 3.3 = 2.32$ in.

$\gamma = 0.59$

Using Equation (6.99):

$C_c = 0.59 \times 5000 \times 0.7 \times 3.3 \times 48 = 327{,}096$ lb

$C_c << T = 449{,}207$ lb

Try a maximum concrete strain of 0.0012 in./in.:

$C = 3.58$ in. $< h_f$

$\beta_1 = 0.705$

$\gamma = 0.63$

$C_c = 381{,}096$ lb $< 449{,}207$ lb

Try a maximum strain of 0.0013 in./in.:

$C = 3.58$ in.

$\beta_1 = 0.71$

$\gamma = 0.66$

$C_c = 433{,}195$ lb

Try $\varepsilon_c = 0.00132$,

$C = 3.9$ in.

$\beta_1 = 0.71$

$\gamma = 0.67$

$C_c = 445{,}255$ lb

$T_f = 0.156 \times 33 \times 10^6 \times 0.0122 = 62{,}806$ lb

$T_s = 1.8 \times 60{,}000 = 108{,}000$ lb

$\varepsilon_{ps} = \dfrac{150{,}000}{27 \times 10^6} + 0.0036 + 0.0122 \left( \dfrac{34.6 - 3.9}{40 - 3.9} \right)$

$\varepsilon_{ps} = 0.0163$ in./in.

$$f_{ps} = 270{,}000 - \frac{75}{0.0163 - 0.0065} - 2000$$

$f_{ps} = 260{,}347$ psi

$T_p = 1.07 \times 260{,}347$

$T_p = 278{,}571$ lb

$T = 62{,}806 + 108{,}000 + 278{,}571 = 449{,}377$ lb

$T = 449{,}377$ lb

$C = 445{,}255$ lb

*Error* < 1%

Use $c = 3.90$ in., $\beta_1 = 0.71$

$\gamma = 0.67$

Nominal moment capacity, (Equation 6.103):

$$M_n = 0.67 \times 0.71 \times 5000 \times 48 \times 3.9 \left(3.9 - \frac{0.71 \times 3.9}{2}\right)$$

$$+ 1.8 \times 60{,}000(38 - 3.9) + 0.85 \times 0.0122 \times 33 \times 10^6$$

$$\times 0.156(40 - 3.9) + 1.07 \times 260{,}347(34.6 - 3.9)$$

$M_n = 15{,}282{,}168$ in.-lb

$\phi M_n = 0.9 \times M_n = 13{,}753{,}951$ in.-lb

Since the original steel area satisfies the minimum reinforcement requirement, only the maximum limit need be checked:

$$\frac{0.85}{d_p} = \frac{0.85 \times 0.71 \times 3.9}{34.6}$$

$$\frac{0.85}{d_p} = 0.068 < 0.36\beta_1 = 0.26.$$

Therefore OK. Hence $M_n = 15{,}282{,}168$ in.-lb

Note that three layers of 0.0065-in. thick and 8-in. wide carbon fiber composite provide an increase of about 13% (for unstrengthened section, $M_n = 13{,}543{,}315$ in.-lb, as calculated in Example 6.5).

## Example 6.10: Unbonded tendons

Repeat Example 6.9, assuming unbonded tendons.

Solution:

Using the results of Example 6.9, the failure will occur by fracture of composite. The contribution of prestressed tendons will be less as compared to bonded tendons. That means that the contribution of the concrete and, hence, the maximum strain in concrete, will be less than 0.00132 in./in. of Example 6.9. Try a maximum concrete strain of 0.0012 in./in. Using the results of Example 6.9,

$$C_c = 381,096 \, \text{lb}$$

For nonprestressed reinforcement:

$$T_s = A_s f_y = 108,000 \, \text{lb}$$

For the composite:

$$T_f = A_f E_f \varepsilon_{fe} = 62,806 \, \text{lb}$$
$$d_p = 34.6 \, \text{in.}$$

$$L = 70 \, \text{ft} = 840 \, \text{in.}, \frac{d_p}{L} = \frac{34.6}{840} = 0.0412$$

Using Equation (6.106), $\Omega_u = 3 \times 0.0412 = 0.124$
Using Equation (6.105):

$$f_{ps} = 150,000 + 0.124 \times 27 \times 10^6 \times 0.0012 \left( \frac{34.6}{3.58} - 1 \right)$$

$$f_{ps} = 184,812 < 0.8 f_{pu} = 216,000 \, \text{psi}$$

Therefore $T_p = 1.07 \times 184,812 = 197,749 \, \text{lb}$

$$T = T_s + T_f + T_p = 368,555 \, \text{lb} < C = 381,096 \, \text{lb}$$

Try maximum concrete strain, $\varepsilon_c = 0.00115$ in./in.: Figure 6.15,

$$c = 40 \left( \frac{0.00115}{0.00115 + 0.0122} \right) = 3.45 \, \text{in.}$$

$$\beta_1 = 0.7$$
$$\gamma = 0.61$$
$$C = C_c = 0.61 \times 0.7 \times 5000 \times 3.45 \times 48$$
$$C = C_c = 353,556 \, \text{lb} < T = 368,555 \, \text{lb}$$

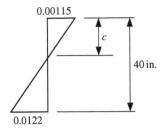

*Figure 6.15* Strain distribution across cross-section for Example 6.10.

The small reduction in depth of the neutral axis will result in an increase of $T$.
Therefore try $\varepsilon_c = 0.00119$ in./in.:

$c = 3.55$ in.

$\beta_1 = 0.7$

$\gamma = 0.62$

$C = 369,768$ lb

$f_{ps} = 184,847$ lb

$T_p = 197,768$ lb

$T = 368,529$ lb $< C = 369,768$ lb

Error $< 0.4\%$. Therefore $c = 3.55$ in.

$$M_n = 369,768 \left( 3.55 - \frac{0.7 \times 3.55}{2} \right) + 108,000(38 - 3.55)$$

$$+ 197,786(34.6 - 3.55) + 0.85 \times 62,806(40 - 3.55)$$

$$M_n = 12,660,991 \text{ in.-lb}$$

Note that for beams with bonded tendons, $M_n$ is 15,282,168 in.-lb (Example 6.9). Check for maximum reinforcement:

$$\frac{0.85a}{d_p} = \frac{0.85 \times 0.7 \times 3.55}{34.6} = 0.061$$

$$\frac{0.85a}{d_p} = 0.061 < 0.36\beta_1 = 0.252.$$

Therefore OK. Hence $M_n = 12,660,991$ in.-lb

## 6.17   Design of strengthening systems: prestressed concrete beams

As in the case of reinforced concrete, the design of strengthening systems involves the following major steps. (Refer to Section 5.22 for more details.)

Estimate the composite area needed for the upgraded loads.

Check the adequacy for factored load (strength analysis).

Check the adequacy for working loads. This step consists of checks for stresses and deflections for short term and long term.

Finalize the details of composite such as the number of layers and width.

The procedures presented in previous sections can be used for the analysis at both working loads and ultimate loads. For estimating the area of composite, the following procedure is recommended.

### 6.17.1   Initial design: estimation of composite area

For the existing section, compute the nominal moment capacity, $M_{ni}$.

For the upgraded loads, estimate the external moment, $M_u$. The composite reinforcement should generate the difference $(M_u - \phi M_{ni})$. In most cases, multiple layers may be required. Depending on the magnitude of the difference, assume the effective failure strain, $\varepsilon_{fe}$, is in the range of 0.5–0.8 $\varepsilon_{fu}$ where $\varepsilon_{fu}$ is manufacturer's guaranteed fracture strain. If the difference, $(M_u - \phi M_{ni})$ is large, use a lower factor. Assuming that the composite will be placed at the extreme tension face, $d_f = h$. The composite area can be estimated using the equation:

$$A_f = \frac{M_u - \phi M_{ni}}{0.9(E_f \varepsilon_{fe} 0.9 d_f)} \tag{6.108}$$

If the repair is carried out to compensate for loss of nonprestressed steel, then $M_u$ is the same. The nominal moment capacity should be calculated without using the lost or ineffective steel area. In this case, the composite area should also satisfy the minimum reinforcement requirements. If the repair is carried out to compensate for the loss of prestress, again, $M_u$ is the same. Moment capacity, $M_n$, should be calculated using the new value for prestress. Again, the combined reinforcement should satisfy the minimum and maximum reinforcement requirements.

In the case of T or double-T beams, the web widths could be small and, hence, the composite may have to be attached at the bottom and sides. In this case, proper value should be used for the effective depth, $d_f$. Once the area, $A_f$, is known, choose the thickness of the plate or the number of layers, and width. Check whether the number of layers used agrees with the reduction factor used for $\varepsilon_{fe}$. If necessary, recompute $A_f$ and the number of layers.

### 6.17.2 Final design

Once the area of composite is determined, analyze the section for working and ultimate loads. For working loads, the allowable stresses are same as the allowable stresses for reinforced concrete. For prestressed tendons at working load, prestress, $f_{ps}$, should be less than $0.74\times$ ultimate stress, $f_{pu}$, and $0.82 \times$ yield stress, $f_{py}$. Minimum yield strength specified by ASTM at a strain of 0.01 in./in. is $0.85f_{pu}$ (ASTM, 2005, 2006).

In addition, the deflection should satisfy the specified requirements.

## 6.18 Design examples: flexural strengthening of prestressed concrete beams

*Example 6.11: Design of strengthening system – bonded tendons*
Design the strengthening system for the following T-beam with bonded tendons, as shown in Figure 6.16.

$f'_c = 5000\,\text{psi}$, $E_c = 4.3 \times 10^6\,\text{psi}$, $f_y = 60{,}000\,\text{psi}$, $E_s = 29 \times 10^6\,\text{psi}$, $E_{ps} = 27 \times 10^6\,\text{psi}$, $f_{pu} = 270{,}000\,\text{psi}$, and $f_{pe} = 150{,}000\,\text{psi}$. Simply supported beam over a span of 70 ft. The repair has to be carried out when dead load of 600 lb/ft is present. The maximum upgraded live load is 600 lb/ft. Use carbon sheets with a guaranteed fracture strain of 0.0167 in./in., modulus of $33 \times 10^6$ psi, and an equivalent fiber thickness of 0.0065 in.

Solution:

Since the cross-section of this example and previous examples on T-section are the same, the results of the previous examples will be used for analysis and design.

$M_{ni} = 13{,}543{,}315$ in.-lb

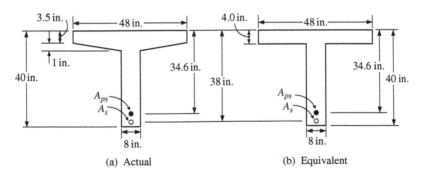

Figure 6.16 Details of cross-section for Example 6.11.

Moment due to factored external loads:

$$M_u = \frac{(1.4 \times 600 + 1.7 \times 600)(70)^2 \times 12}{8}$$

$$M_u = 13,671,000 \text{ in.-lb}$$

$$\frac{M_u}{\phi} = \frac{M_u}{0.9} = 15,190,000 \text{ in.-lb}$$

Area of composite:

$$A_f = \frac{\left(\frac{M_u}{\phi} - M_{ni}\right)}{0.9(E_f \varepsilon_{fe} 0.9 d_f)}$$

Assume $\varepsilon_{fe} = 0.8 \varepsilon_{fu}$,

$d_f = 40 \text{ in.}$

$$A_f = \frac{15,190,000 - 13,543,315}{0.9 \times 33 \times 10^6 \times 0.8 \times 0.0167 \times 0.9 \times 40}$$

$$A_f = 0.115 \text{ in.}^2$$

For a thickness of 0.0065 in., width = 18 in., try three layers, 8-in. wide:

Therefore $A_f = 3 \times 8 \times 0.0065 = 0.156 \text{ in.}^2$

Check for ultimate load:
Using the results of Example 6.9:

$\varepsilon_{bi} = 0$

Nominal moment capacity for beam with three layers of carbon:

$$M_n = 15,282,168 \text{ in.-lb} > \frac{M_u}{\phi} = 15,190,000 \text{ in.-lb.}$$

Hence OK. Note that both minimum and maximum reinforcement ratios are satisfied.
Check for allowable stresses at working load:

Maximum working load moment, $\dfrac{(600 + 600)(70)^2 \times 12}{8} = 8,820,000$ in.-lb

Using the results of Example 6.7, in which the analysis was done for a moment of 9,000,000 in.-lb:

$f_{ct} = 1505 \text{ psi} < 0.45 f_c' = 2250 \text{ psi}$

$f_s = 38,076 \text{ psi} < 0.8 f_y = 48,000 \text{ psi}$

$f_{ps} = 181,451 \text{ psi} < 0.74 f_{pu} = 199,800 \text{ psi}$

If $f_{py}$ is assumed $0.85 f_{pu}$,

$f_{ps}$ is less than $0.82 f_{py} = 0.82 \times 0.85 f_{pu} = 188,190 \text{ psi}$

Fiber stress, $f_f = 46,200 \text{ psi} < 0.33 \times 0.9 E_f \varepsilon_{fu} = 163,676 \text{ psi}$

Therefore the composite satisfies the allowable stress requirement.

## *Example 6.12: Design of strengthening system for repair*

For the beam cross-section shown in Figure 6.6 (Example 6.2), assume that one of the tendons is lost due to an accident. Design a repair system to restore the strength of the T-beam. Use carbon fibers with $\varepsilon_{fu} = 0.0167$, $E_f = 33 \times 10^6 \text{ psi}$, and equivalent fiber thickness of 0.0065 in.

Solution:

Using the given details and computations of Examples 6.2 and 6.5, the following are known:

$b = 48 \text{ in.}$

$b_w = 8 \text{ in.}$

$d_p = 34.6 \text{ in.}$

$d_s = 38 \text{ in.}$

$h = 40 \text{ in.}$

$A_s = 1.8 \text{ in.}^2$

Original $A_{ps} = 1.07 \text{ in.}^2$ (7 1/2-in. strands)
For 6 1/2-in. strands, $A_{ps} = 0.917 \text{ in.}^2$

$f_c' = 5000 \text{ psi}$

$f_y = 60,000 \text{ psi}$

$f_{pu} = 270,000 \text{ psi}$

For seven strands, $M_n = 13,543,315 \text{ in.-lb}$ (Example 6.5).
Computation of nominal moment capacity for six strands:
Assuming depth of stress block, $a$ is $< h_f$:

$0.85 \times 5000 \times 48 \times a = A_s f_y + A_{ps} f_{ps}$

$A_s, f_y,$ and $A_{ps}$ are known. For $f_{ps}$,

$\gamma_p = 0.28, \beta_1 = 0.8$

$$f_{ps} = 270,000 \left[ 1 - \frac{0.28}{0.8} \left\{ \left( \frac{0.917}{48 \times 34.6} \times \frac{270,000}{5000} \right) \right. \right.$$

$$\left. \left. + \left( \frac{38}{34.6} \times \frac{1.8}{48 \times 38} \times \frac{60,000}{5000} \right) \right\} \right]$$

$f_{ps} = 265,953 \, \text{psi}$

Therefore $0.85 \times 5000 \times 48 \times a = 1.8 \times 60,000 + 0.917 \times 265,953$

$a = 1.73 \, \text{in.} < h_f$. Therefore OK.

$$M_n = 0.85 \times 5,000 \times 48 \times 1.73 \left( 34.6 - \frac{1.73}{2} \right) + 1.8$$

$$\times 60,000 \, (38 - 34.6)$$

$M_n = 12,272,956 \, \text{in.-lb}$

This value is going to be initial moment capacity, $M_{ni}$.
Estimation of $A_f$ required:
Assuming $\varepsilon_{fe} = 0.8\varepsilon_{fu}$ and $d_f = 40 \, \text{in.}$:

$$A_f = \frac{13,543,315 - 12,272,96}{0.9 \times 33 \times 10^6 \times 0.8 \times 0.0167 \times 0.9 \times 40}$$

$A_f = 0.09 \, \text{in.}^2$

For a fiber thickness of $0.0065 \, \text{in.}$, width $\cong 14 \, \text{in.}$
Try two layers of 8-in. wide strips,

$$A_f = 16 \times 0.0065 = 0.104 \, \text{in.}^2$$

Assume that the repair is only with dead load and $\varepsilon_{bi} \cong 0$. Note that the force lost to one tendon should be compensated by $A_f$. Assuming $f_{ps}$ of $266,000 \, \text{psi}$:

Force to be replaced, $\cong 0.0152 \times 266,000 = 40,432 \, \text{lb.}$

For $A_f = 0.104 \, \text{in.}^2$, stress at ultimate load $= \dfrac{40,432}{0.104} = 388,769 \, \text{psi}$

$$\varepsilon_f = \frac{388,769}{33 \times 10^6} = 0.0118 \, \text{in./in.}$$

This strain could be considered reasonable.
Computation of the ultimate moment for the strengthened section:

Based on the results of Example 6.9, assume that failure will be initiated by failure of the composite.
Compute effective strain for composite, $\varepsilon_{fe}$:
For two layers,

$$ntE_f = 2 \times 0.0065 \times 33 \times 10^6 = 429,000$$

$$\text{Therefore } \varepsilon_{fe} = \left(1 - \frac{429,000}{2,400,000}\right) 0.0167 = 0.0137$$

$$\varepsilon_{fe} = 0.0137 < 0.90 \times 0.0167$$

Try a maximum concrete strain of 0.0013 in./in. at failure, as shown in Figure 6.17,

$$\frac{c}{0.0013} = \frac{40 - c}{0.0137}$$

$$c = 3.47 \text{ in.}$$

Using nonlinear analysis, $\gamma = 0.66$

$$\beta_1 = 0.71$$

$$C_c = 0.710 \times 0.66 \times 48 \times 3.47 \times 5000 = 390,250 \text{ lb}$$

$$T = 1.8 \times 60,000 + 0.0137 \times 33 \times 10^6 \times 0.104 + 0.917 f_{ps}$$

$$\varepsilon_{ps} = \frac{150,000}{27 \times 10^6} + \varepsilon_{ce} + 0.0137 \left(\frac{34.6 - 3.47}{40 - 3.47}\right)$$

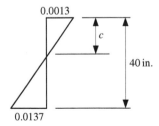

*Figure 6.17* Strain distribution for Example 6.12.

Assume $\varepsilon_{ce} = 0.00036$ from Example 6.9. The value will be slightly smaller because one tendon is lost. However, the influence of $\varepsilon_{ce}$ on $\varepsilon_{ps}$ is negligible.

$\varepsilon_{ps} = 0.0176$ in./in.

$$f_{ps} = 270,000 - \frac{75}{0.0176 - 0.0065} - 2000$$

$f_{ps} = 261,000$ psi

$T = 394,300$ lb

$C_c = 390,250$ lb. Error $< 1\%$

Maximum strain smaller than 0.0013 in./in. Since the error is small, compute $M_n$ using $\varepsilon_c = 0.0013$ in./in.

$$M_n = 390,250 \left( 3.47 - \frac{0.71 \times 3.47}{2} \right) + 1.8 \times 60,000(38 - 3.47)$$

$$+ 0.917 \times 261,000(34.6 - 3.47) + [0.85 \times 0.104 \times 0.0137 \times 33$$

$$\times 10^6 (40 - 3.47)]$$

$M_n = 13,513,184$ in.-lb

This moment is 0.22% less than $M_n$ of original section, 13,543,315 in.-lb. However, $\frac{M_u}{\phi}$ required for dead load of 613 lb/ft and live load of 500 lb/ft is:

$$\frac{M_u}{\phi} = 12,561,973 < M_n = 13,513,184 \text{ in.-lb}$$

Therefore the design is satisfactory for ultimate load. Note that the reinforcement satisfies the requirements of minimum and maximum reinforcement.

Since the working load stresses calculated in Example 6.7 are well within allowable limits, it is assumed that two layers of composite are satisfactory.

## 6.19    Problems

**Problem 6.1:** Compute the cracking moment for the following cross-sections. Assume that the tendons are bonded. $E_s = 29 \times 10^6$ psi, $E_{ps} = 27 \times 10^6$ psi, and $E_f = 33 \times 10^6$ psi.

(a)  rectangular section, Figure 6.18;
(b)  double-tee section, Figure 6.19;
(c)  AASHTO – PCI Bulb-Tee, Figure 6.20.

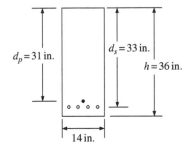

*Figure 6.18* Cross-section of rectangular beam for Problem 6.1(a).
*Notes*: Span = 30 ft; $f'_c$ = 5000 psi; $A_s$ – 4 No. 7 bars; $A_{ps}$ – 5 – 7/16 in. 7 wire strands; $f_y$ = 60,000 psi; Cable depressed at center span.

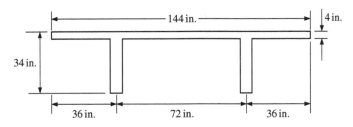

*Figure 6.19* Cross-section of double-tee beam for Problem 6.1(b).
*Notes*: Span = 60 ft; $A$ = 978 in.²; $I$ = 86,072 in.⁴; $y_t$ = 8.23 in.; $w_t$ = 1019 lbs/ft; $f'_c$ = 5000 psi; $f_{pu}$ = 270,000 psi; $f_y$ = 60,000 psi; $A_s$ – 4 No. 8 bars; $A_{ps}$ – 16 – 0.5 in. diameter strands; $e_{center}$ = 2.2 in.; $e_{support}$ = 12.75 in.; Cable depressed at center span.

**Problem 6.2:** For the cross-sections of Problem 6.1, compute the maximum stresses in concrete, nonprestressed, and prestressed steel for a maximum moment of $1.5M_{cr}$. Compute the deflection for this moment.

(i)  Assume bonded tendons.
(ii) Assume unbonded tendons.

**Problem 6.3:** For the cross-sections of Problem 6.1, compute the nominal moment capacity, $M_n$ assuming (a) bonded, and (b) unbonded tendons. If the dead load is 1.4 times self-weight, estimate the safe maximum live load that can be carried by these beams. Assume that all the loads are uniformly distributed.

**Problem 6.4:** Repeat Problems 6.1 and 6.2 for strengthened cross-sections. Guaranteed $\varepsilon_{fu}$ = 0.0176 in./in. Equivalent fiber thickness = 0.0065 in. Only dead load was present during strengthening.

For the rectangular section, two layers 4- in. wide each were bonded to the bottom surface.

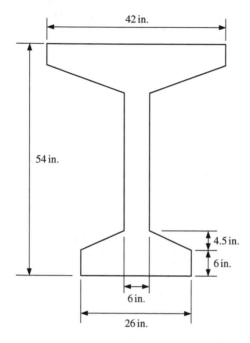

*Figure 6.20* Cross-section of PCI Bulb-Tee beam for Problem 6.3(c).
*Notes:* Span $= 100$ ft; $A = 659$ in.$^2$; $I = 268,077$ in.$^4$; $y_b = 27.63$ in.; $w_t = 686$ lbs/ft; $f'_c = 7000$ psi; $f_{pu} = 270,000$ psi; $f_y = 60,000$ psi; $A_s - 10$ No. 8 bars; $A_{ps} - 50 - 0.5$ in. diameter cables; $e_{center} = 40$ in.; Assume straight cables.

For the Double-T, three layers that are 10 in. wide each were bonded to the bottom of the webs.

For the I-section, three layers that are 26 in. wide each were bonded to the bottom flange.

**Problem 6.5:** Design the strengthening system if the moment capacity computed in Problem 6.3 has to be increased by 20% for cases (a)–(c).

## References

AASHTO Subcommittee on Bridges and Structures (1975) *Interim Specifications: Bridges.* Washington, DC: American Association of State Highway Transportation Officials.

ACI Committee 318 (2005) *Building Code Requirements for Structural Concrete and Commentary (ACI 318-05).* Farmington Hills, MI: American Concrete Institute.

ACI-ASCE Joint Committee 423 (2005) *Recommendations for Concrete Members Prestressed with Unbonded Tendons (ACI 423.3R-05).* Farmington Hills, MI: American Concrete Institute.

American Society for Testing and Materials (2005) *ASTM Standard A421: Standard Specification for Uncoated Stress-Relieved Steel Wire for Prestressed Concrete.* West Conshohocken, Pennsylvania: ASTM.

American Society for Testing and Materials (2006) *ASTM Standard A416: Standard Specification for Steel Strand, Uncoated Seven-Wire for Prestressed Concrete.* West Conshohocken, Pennsylvania: ASTM.

Balaguru, P.N. (1981) Increase of stress in unbonded tendons in prestressed concrete beams and slabs. *Canadian Journal of Civil Engineering*, 8(2), pp. 262–268.

Cha, J.Y., Balaguru, P.N., and Chung, L. (1999) Experimental and analytical investigation of partially prestressed concrete beams strengthened with carbon reinforcement. In: Dolan, C.W., Rizkalla, S.H., and Nanni, A. (eds), *Proceedings of the 4th International Symposium on Fiber Reinforced Polymer Reinforcement for Reinforced Concrete Structures (FRPRCS-4)*, November. Baltimore, MD, pp. 625–633.

Collins, M.P. and Mitchell, D. (2005) *Reinforced and Prestressed Concrete Structures.* 2nd edn. London: Spon Press, 784 pp.

Lin, T.Y. and Burns, N.H. (1981) *Design of Prestressed Concrete Structures.* 3rd edn. New York: John Wiley & Sons, 646 pp.

Naaman, A. (2004) *Prestressed Concrete Analysis and Design: Fundamentals.* 2nd edn. Ann Arbor, Michigan: Techno Press 3000, 1072 pp.

Nawy, E.G. (2003) *Prestressed Concrete: A Fundamental Approach.* 4th edn. Upper Saddle River, NJ: Pearson Prentice Hall, 939 pp.

Nilson, A.H. (1987) *Design of Prestressed Concrete.* 2nd edn. New York: John Wiley & Sons, 608 pp.

Post-Tensioning Institute (1999) *Post-Tensioning Manual.* 5th edn. Phoenix, AZ: PTI.

Precast/Prestressed Concrete Institute (1999) *PCI Design Handbook: Precast and Prestressed Concrete.* 6th edn. Chicago: Precast/Prestressed Concrete Institute.

Rosenboom, O., Hassan, T.K., and Rizkalla, S. (2004) Flexural behavior of aged prestressed concrete girders strengthened with various FRP systems. *Construction and Building Materials*, 21(4), pp. 764–776.

Rosenboom, O., Walter, C., and Rizkalla, S. (2008) Strengthening of prestressed concrete girders with composites: Installation, design and inspection. *Construction and Building Materials*, doi: 10.1016/j.conbuildmat.2007.11.010. Available online 10 January 2008.

# 7    Shear in beams

## 7.1 Introduction

High-strength composites have been evaluated for strengthening in shear mode by a number of investigators (Anil, 2006; Mosallam and Banerjee, 2007; Triantafillou, 1998). A number of field applications have also been carried out all over the world (Kachlakev and McCurry, 2000; Seible, 1995). These include reinforced and prestressed concrete beams and columns. This chapter deals with the beams while the application to columns is presented in Chapter 8.

The composites are applied in the form of sheets, plates, or bars along the depth of the beam or perpendicular to the potential shear cracks. Shear strengthening was also found to improve the ductility because of the partial confining provided by the strengthening systems. The chapter presents a summary of the provisions used for reinforced and prestressed concrete, the schemes used for composite general guidelines, strain and stress limits, design procedures, and examples (ACI Committee 318, 2005; ACI Committee 440, 2002). As in the case of other chapters, the reader is referred to texts on reinforced concrete for detailed discussions on mechanisms and provisions (Limbrunner and Aghayere, 2007; MacGregor and Wight, 2005; Nawy, 2005; Setareh and Darvas, 2007; Wang *et al.*, 2007).

## 7.2 Failure mechanism of reinforced concrete beams

In the case of flexural failure, the flexural cracks at the maximum bending moment location move toward the compression zone creating a hinge before failure. The curvature and the rotation at this location result in excessive deflection, providing a warning of impending failure. In the case of shear, the failure could occur due to diagonal tension or shear compression failure, and in both cases, the failure is much more brittle than the flexural failure. These failure modes are briefly discussed below.

### 7.2.1   Diagonal tension failure

This type of failure occurs in beams with a shear span ratio in the range of 2.5–5.5 (Figure 7.1). Shear span is designed as the distance between the load and the support (Figure 7.1). As the load increases, the region near the support where shear stresses are predominant, develops diagonal stresses that could exceed the tensile strength of concrete. In some instances, small flexural

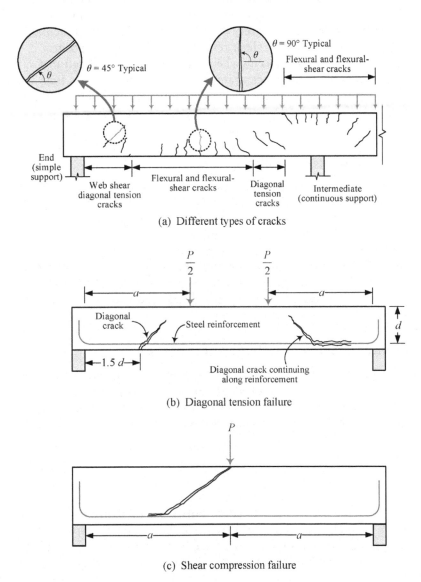

(a) Different types of cracks

(b) Diagonal tension failure

(c) Shear compression failure

*Figure 7.1* Crack patterns and shear failure modes (Source: ACI Committee 440, 2002; Nawy, 2003).

cracks that are perpendicular to the axis of the beam could join the diagonal cracks. In some instances, the crack propagates along the reinforcement and could result in pull-out of the tension bar if the bar is not properly anchored. The critical location for diagonal cracks is between 1.5 and 2.5$d$ from the support. The diagonal cracks, when occur at 45° to the axis of the beam, cover a distance of $d$ along the axis. Owing to the presence of normal (flexural) stresses, the angles of cracks do not exceed 45°. This is one of the reasons for limiting the maximum spacing of stirrups to $d$. This limit assures the presence of reinforcement at every diagonal crack. The same principle should be followed when designing shear reinforcement using composites.

Diagonal shear cracks occur when the shear span-to-depth $a/d$ ratio is between 2.5 and 5.5 for concentrated loads. If the ratio is less than 2.5, the mode of failure changes to shear–compression. This is discussed in the next section.

### 7.2.2   Shear compression failure

This type of failure occurs when the $a/d$ ratio is less than 2.5 for concentrated loads, and 5 for distributed loads. These beams are typically referred to as deep beams. In this type of beam, the crack essentially travels from the load point to a location near the support (Figure 7.1(c)). The propagation is a result of both flexural and shear stresses. Deep beams are excellent candidates for the use of composite because they provide a large surface area. A system with fibers oriented in mutually perpendicular directions will provide an efficient reinforcement. The repair will also provide some confinement.

## 7.3   Shear strength contribution of concrete

The external shear force is resisted by concrete and reinforcement. The contribution of concrete is assumed proportional to square root of concrete strength because the tensile strength of concrete is proportional to $\sqrt{f'_c}$. The contribution of concrete to shear resistance can be conservatively estimated using,

$$V_c = 2\sqrt{f'_c}b_w d \tag{7.1}$$

where $b_w$ is the web width. If sand-lightweight or lightweight concrete is used, $V_c$ has to be multiplied by a factor of 0.85 or 0.75, respectively.

The presence of tension steel used for flexure influences the shear behavior of concrete. Extensive investigations conducted around the world showed that the shear contribution of concrete in the presence of reinforcement is influenced by the amount of web reinforcement $\rho_w$, and the ratio, $V_u/M_u$. Note that $\rho_w$ is the web reinforcement ratio (equal to $A_s/b_w d$), while $V_u$

and $M_u$ are ultimate shear and moment values, respectively. Based on the recommendation of ACI code, the contribution of concrete can be written as:

$$V_c = 1.9 b_w d \sqrt{f_c'} + 2500 \rho_w \frac{V_u d}{M_u} b_w d \leq 3.5 b_w d \sqrt{f_c'} \tag{7.2}$$

Neither $\dfrac{V_u d}{M_u}$ nor $\dfrac{V_n d}{M_n}$ can exceed 1.0.

The values of $V_u$ and $M_u$ are computed at a distance $d$ from the support. Note that if nominal values are used, the ratio of $V/M$ might be slightly different. The shear capacity can also be computed using Equation (7.1), even though the values are typically smaller than the values calculated using Equation (7.2).

If axial compression is present:

$$V_c = 2 \left( 1 + \frac{N_u}{2000 \, A_g} \right) b_w d \sqrt{f_c'} \tag{7.3}$$

where $A_g$ is the gross area measured in units of in.$^2$, while $N_u$ is the axial force in pounds. For axial tension:

$$V_c = 2 \left( 1 + \frac{N_u}{500 \, A_g} \right) b_w d \sqrt{f_c'} \tag{7.4}$$

In Equation (7.4), $N_u$ is negative.

For circular members, the area $b_w d$ is taken as the product of the diameter and the effective depth. The effective depth can be taken as 0.8 times the diameter.

## 7.4 Shear strength contribution of steel

The shear strength reinforcement performs three basic functions:

1. carries a portion of the external shear force;
2. restricts the growth of diagonal cracks;
3. holds the flexural reinforcement bars in place so that they can provide dowel action.

If stirrups are provided as closed ties, they also provide confinement to the concrete in the compression zone. A number of models are available for estimating the contribution of stirrups to shear capacity. The popular models are truss analogy and compression strut theory. Essentially, the compressive forces are taken by the concrete and the steel provides the tensile force needed for equilibrium. If the steel stirrups with area, $A_v$, are provided at a spacing of $s$, then the contribution of steel:

$$V_s = \left(\frac{A_s f_y d}{s}\right)(\sin \beta + \cos \beta) \tag{7.5}$$

where $f_y$ is the yield strength of steel used for the stirrups and $\beta$ is the angle of the stirrups relative to the axis of the beam. If the stirrups are vertical, Equation (7.5) becomes:

$$V_s = \left(\frac{A_s f_y d}{s}\right) \tag{7.6}$$

This equation can be used for computing both the contribution of the steel stirrups and the stirrup spacing.

## 7.5   Limitations on spacing and maximum contribution of reinforcement

In order to ensure that stirrups are present at all diagonal cracks, the following restrictions are placed on stirrup spacing. The spacing cannot be less than 24 in.:

$$\text{If } V_n - V_c \le b_w d \sqrt{f_c'}, s_{max} = \frac{d}{2} \tag{7.7}$$

$$\text{If } V_n - V_c > b_w d \sqrt{f_c'}, s_{max} = \frac{d}{4} \tag{7.8}$$

The contribution of steel should not exceed $8 b_w d \sqrt{f_c'}$. If the situation occurs, the section has to be enlarged. If the factored shear force, $V_u$, exceeds half the capacity of the plain concrete section, stirrups should be provided. In addition:

$$A_v \ge \frac{50 b_w s}{f_y} \tag{7.9}$$

where $A_v$ is the area of all the vertical legs. The last two provisions are provided as a precaution against brittle failure.

## 7.6   Composites for shear strengthening

A number of investigators have shown that composites can be effectively used to increase the shear strength of beams and columns (Chajes *et al.*, 1995; Khalifa *et al.*, 1998; Kachlakev and McCurry, 2000; Malvar *et al.*,

1995; Norris *et al.*, 1997). A number of configurations can be used for shear strengthening. These configurations are briefly discussed in the following sections.

### 7.6.1 *Sheets bonded to web*

The composite can be placed on the sides of the web as shown in Figure 7.2. The most effective way is to wrap the entire cross-section as shown in Figure 7.2. Even though, theoretically, this is the best option, it may not be practical in most cases because taking the composite through the slab will be very expensive. In most cases, it will also be impractical because of the floor covering on top of the slab.

A more practical approach is to cover the entire web as shown in Figure 7.2(b). This type of wrap, referred to as "U-wrap," provides good anchor. However, these wraps are not as effective in the negative moment region because shear cracks initiate at top of the beam. For positive moment regions, U-wraps are the most commonly used configurations.

Providing a mechanical anchor at the junction of the beam and slab can considerably increase the effectiveness of U-wraps. This anchor is provided by cutting a groove and placing the sheets in the groove with the help of a bar. This procedure is shown to increase the force-carrying capacity of wrap considerably. In some cases, the bottom beam surface may not be accessible. In these cases, the composite can be applied only to the sides (Figure 7.2(c)). In terms of shear strength contribution, this is the least effective. However, in some cases this might be the only choice. If composite bars or plates are used for strengthening, this configuration is the only choice. If bars are used, grooves need to be cut so that those bars can be embedded inside to provide sufficient bond strength.

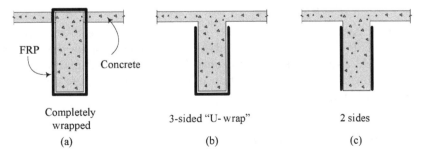

*Figure 7.2* Various schemes for wrapping transverse FRP reinforcement (Source: ACI Committee 440, 2002).

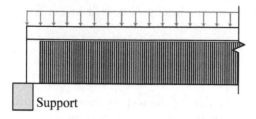

(a) Continuous reinforcement

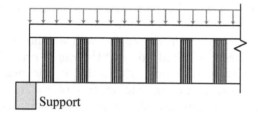

(b) Reinforcement strips

*Figure 7.3* Shear reinforcement distributions.

### 7.6.2 Spacing of reinforcement

The composite can be applied as a continuous sheet, or strips can be placed at a designed spacing (Figure 7.3). When prefabricated plates or bars are used, the second option is the only choice (Figure 7.3(b)). Continuous wrap might provide better confinement, but may not allow vapor release. If it is suspected that moisture could migrate into concrete through uncovered areas, the second option should be used to avoid degradation at the concrete composite interface.

### 7.6.3 Fiber orientation

The fiber can be oriented vertically or at a chosen angle to maximize the fiber contribution (Figure 7.4(a) and (b)). This can be done for sheets applied in the field and prefabricated systems. For prefabricated plates and bars, application in the inclined direction might be better because there will be more bond length.

### 7.6.4 Biaxial reinforcement

Fibers can be placed in the mutually perpendicular directions as shown in Figure 7.5. Typically, they will be placed in horizontal and vertical directions

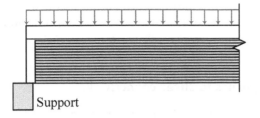

(a) Wrap using $\theta = 90°$

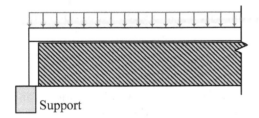

(b) Wrap using $\theta = 45°$

*Figure 7.4* FRP reinforcement sheets oriented in various primary directions.

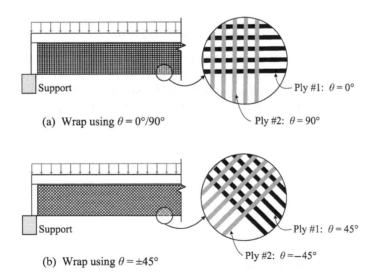

(a) Wrap using $\theta = 0°/90°$

Ply #1: $\theta = 0°$

Ply #2: $\theta = 90°$

(b) Wrap using $\theta = \pm 45°$

Ply #1: $\theta = 45°$

Ply #2: $\theta = -45°$

*Figure 7.5* Beams reinforced with biaxial FRP shear reinforcement.

(Figure 7.5(a)). The fibers in mutually perpendicular directions act both as reinforcement and as anchors; they are found to provide a synergistic effect. If prefabricated plates or bars are used, this may not be an option, unless very thin plates are used.

## 7.7   Contribution of composites to shear capacity, $V_f$

The principles that apply to steel stirrups also apply to composite reinforcement. The major differences are the force transfer mechanism and the high strength of the composite. The force is transferred through shear at the interface or via mechanical anchors. The failure stress of the composite could be much higher than the yield strength of steel, $f_y$. Once $V_f$ is known, it can be added to the contributions of concrete, $V_c$, and steel, $V_s$, to obtain the total shear capacity, $V_n$:

$$V_n = V_c + V_s + V_f \qquad (7.10)$$

The contribution of composite, $V_f$ depends on several parameters including the stiffness of the composite, thickness of the composite, compressive strength of concrete, fiber orientation angle, quality of resin used for bonding, and anchoring mechanisms. The failure of the composite could occur either by composite rupture or by delamination. In both cases, failure will occur well before the composite can reach the fracture stress obtained in tension tests.

### 7.7.1   *Shear contribution based on composite rupture*

The design approach is based on fracture of composite, and is similar to the approach used for steel reinforcement. Instead of yield stress $f_y$ for steel, effective fracture stress, $f_{fe}$, is used for the computation. Triantafallou (1998) documented the reduction in fracture stress that occurs due to stress concentrations. If the effective strain at failure is $\varepsilon_{fe}$, then:

$$f_{fe} = E_f \varepsilon_{fe} \qquad (7.11)$$

and

$$V_f = \frac{A_{fv} E_f \varepsilon_{fe} (\sin \beta + \cos \beta) d_{fv}}{s_f} \qquad (7.12)$$

where $A_{fv}$ is the area of the composite reinforcement, $d_{fv}$ is the depth to the composite, and $s_f$ is the spacing of the composite reinforcement (Figure 7.6). Note the similarities between Equations (7.5) and (7.12). $A_v$ is replaced with $A_{fv}$, $d$ is replaced with $d_{fv}$, $s$ is replaced with $s_f$, and $f_y$ is replaced with $f_{fe}$

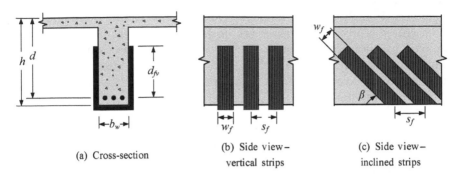

(a) Cross-section

(b) Side view – vertical strips

(c) Side view – inclined strips

*Figure 7.6* Variables and dimensions used to define the area of FRP for shear (Source: ACI Committee 440, 2002).

or $E_f \varepsilon_{fe}$. The area of composite is $2t_f$ times $w_f$ where $t_f$ is the thickness of composite placed on both sides and $w_f$ is the width of composite. If the composite is continuous, then the width of the strip, $w_f$, becomes equal to the spacing, $w_f$, or $A_f/s_f$ becomes twice the thickness of the composite. As in the case of flexure, it is recommended to use only the fiber area of equivalent fiber thickness and fiber modulus. Typically, composite strips extend only to the soffit of the slab at the top of the beam (Figure 7.6). Therefore the effective depth, $d_{fv}$, should be taken as $(d - h_f)$, where $d$ is the effective depth to flexural steel and $h_f$ is the flange thickness. This is done to ensure that composite shear reinforcement is present in the shear cracks developed in the web between steel reinforcement level and the soffit of the slab. For rectangular beams, $d_{fv}$ can be taken as $d$.

The angle $\beta$ is determined by the orientation of composite (Figure 7.6). The reinforcement is most effective when placed perpendicular to the potential shear crack. In most cases, composite strips are placed in the vertical direction for ease of application and less material use. However, if the potential crack direction is known, or longer bond length is needed, inclined strips can be used.

The final parameter in Equation (7.12) is $E_f \varepsilon_{fe}$. Since $E_f$ is a known material property, the only unknown is the effective fiber strain at failure, $\varepsilon_{fe}$. If the composite is wrapped around or anchored at the soffit of the slab, the probability for fiber fracture is high. At large crack widths, it was found that aggregate interlock is lost (Priestley and Seible, 1995; Priestley *et al.*, 1996). To avoid this, the maximum effective strain is limited to the minimum of 0.004 or $0.75\varepsilon_{fu}$ (guaranteed fracture strain). In other words, for systems with complete wrap or anchored systems guaranteed to fracture the composite:

$$\varepsilon_{fe} = 0.004 \le 0.75\varepsilon_{fu} \tag{7.13}$$

where $\varepsilon_{fe}$ is the guaranteed fracture strain provided by the composite manufacturer. If analysis that is more sophisticated is warranted, say in a large rehabilitation project, the procedure outlined by Khalifa *et al.* (1998) can be used for the estimation of $\varepsilon_{fe}$. The results reported in this paper confirm that, for the composites wrapped around the beam, the effective strain will be equal to or exceed 0.004. If the thickness of composite is very small, the effective strains are higher.

### 7.7.2　Shear contributions based on delamination

Delamination is not only a much weaker mechanism but also susceptible to a large number of variables. The contribution can still be estimated using Equation (7.12), but the estimation of effective strain, $\varepsilon_{fe}$ is different. For bond failure or delamination mechanism, the major variables are effective bond length, $L_e$, thickness of the composite, the compressive strength of concrete, and the type of wrapping scheme. The effective bond length, $L_e$, decreases with increasing stiffness of the composite. For a given fiber type, increase in thickness leads to decrease in effective bond length. Based on experimental results reported by various authors, ACI committee 440 (2002) recommends the following equations for estimating $L_e$ in inch-pound units:

$$L_e = \frac{23,000}{\left(n_f\, t_f E_f\right)^{0.58}} \tag{7.14}$$

where $n\,t_f$ is again equivalent unit thickness. If fiber sheets are used, the equivalent fiber thickness is multiplied by the number of plies. For prefabricated plate, the equivalent fiber thickness of the plate should be used. To account for the difference in compressive strength, a factor $k_1$ is used. For inch-pound units:

$$k_1 = \left(\frac{f'_c}{4000}\right)^{\frac{2}{3}} \tag{7.15}$$

Note that the variation is standardized with respect to concrete with a compressive strength of 4000 psi. Coefficients of U- and two-sided bonded wraps are as follows:

$$k_2 = \frac{d_{fv} - L_e}{d_{fv}} \text{ for U-wrap} \tag{7.16}$$

$$k_2 = \frac{d_{fv} - 2L_e}{d_{fv}} \text{ for two-sided bonded} \tag{7.17}$$

For other types such as bars placed in grooves, the experimental results and the recommendations of the authors of the investigation should be used for estimating $k_2$. Equations (7.16) and (7.17) are based on results obtained using simply supported beams representing high shear and low moment. This may not be applicable to locations near intermediate supports of continuous beams where both high shear and high moment exist. For these cases, it is recommended to use a more conservative value. Combining the various effects, fracture strain can be obtained using a reduction coefficient, $\kappa_V$, which is similar to the coefficient used for flexure:

$$\varepsilon_{fe} = \kappa_V \varepsilon_{fu} \leq 0.004 \text{ for U-wraps for bonding on two sides} \tag{7.18}$$

$$\kappa_V = \frac{k_1 k_2 L_e}{468\, \varepsilon_{fu}} \leq 0.75 \tag{7.19}$$

$L_e$, $k_1$, and $k_2$ are estimated using Equations (7.14), (7.15), and (7.16) or (7.17), respectively. In summary, the contribution of composite $V_f$ to shear capacity is computed using Equations (7.12), (7.13), and (7.18).

## 7.8  Limitations on total shear reinforcement, spacing, and strength reduction factor

Following the general philosophy of ACI code, the total contribution of reinforcement should be limited. Using the principle of steel reinforcement:

$$V_s + V_f \leq 8\sqrt{f_c'}b_w d \tag{7.20}$$

When computing total shear capacity, the contribution of the composite is reduced by a factor $\psi_f$. Based on current knowledge, factors of 0.95 and 0.85 are recommended for completely wrapped and U-plies, respectively (ACI Committee 440, 2002). Therefore, shear capacity of the section, computed using:

$$\phi\, V_n = \phi\,(V_c + V_s + \psi_f V_f) \tag{7.21}$$

should be equal to or greater than maximum shear force calculated using factored loads, or:

$$\phi\, V_n \geq V_u \tag{7.22}$$

Similarly to steel reinforcement, spacing of composite reinforcement should be restricted to make sure that reinforcement is present at potential shear cracks. Taking a conservative approach:

$$s_f \leq w_f + \frac{d_{fv}}{4} \tag{7.23}$$

## 7.9  Design procedure

The following is the recommended design procedure, presented in a sequence of steps.

*Step 1*: Determine the critical section, and compute the maximum shear force for the factored loads, $V_u$. The critical section is at a distance, $d$, from the support provided that the loads introduce compression into the end region of the member and there are no concentrated loads between the support and distance, $d$.

*Step 2*: Compute $V_c$ and check whether

$$\frac{V_u}{\phi} - V_c \leq 8\sqrt{f_c'}b_w d \tag{7.24}$$

If this condition is not satisfied, the section needs enlargement since composites are used to enhance the existing structure. If heavy shear reinforcement is provided in the original design, the condition might control the design.

*Step 3*: Compute the shear contribution of existing steel reinforcement,

$$V_s = \frac{A_v f_y d}{s}(\sin \beta + \cos \beta) \tag{7.25}$$

For vertical stirrups, $(\sin \beta + \cos \beta) = 1$

*Step 4*: Estimate the shear to be carried by composite, $V_f$:

$$V_f = \frac{1}{\psi_f}\left[\frac{V_u}{\phi} - V_c - V_s\right] \tag{7.26}$$

$\psi_f$ is 0.85 and 0.95 for two-sided and three-sided wraps, respectively.

*Step 5*: Based on the system selected, compute $\varepsilon_{fe}$ and spacing, $s_f$, using,

$$s_f = \frac{A_{fv}E_f\varepsilon_{fe}(\sin \beta + \cos \beta)d_{fv}}{V_f} < w_f + \frac{d_{fv}}{4} \leq 24\,\text{in.} \tag{7.27}$$

Again, if the composite is placed in vertical strips, $(\sin \beta + \cos \beta) = 1$. These steps are presented in the flow chart (Figure 7.7).

### Example 7.1: Composite shear reinforcement for T-beam

The T-beam shown in Figure 7.8 is simply supported over a span of 30 feet and supports a uniformly distributed dead load of 1300 lb/ft and a live load

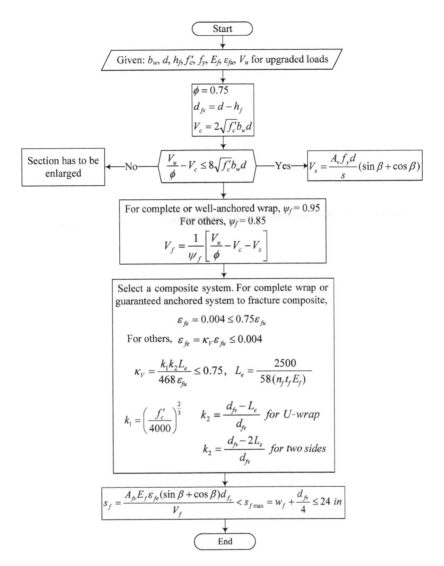

*Figure 7.7* Flowchart for design of shear reinforcement.

of 1600 lb/ft. In addition, there is a concentrated load of 5000 lb acting at mid-span. Stirrups, designed for the old set of loads, consist of ASTM No. 3 two-legged stirrups spaced at 12 in. center-to-center. Verify the adequacy of the stirrups and design the extra shear reinforcement, if warranted. The compressive strength of concrete and the yield strength of stirrup steel was 4000 and 40,000 psi, respectively.

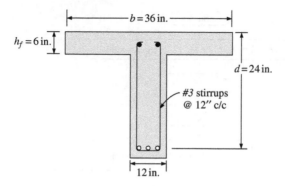

*Figure 7.8* Details of beam for Example 7.1.

Solution:

For distributed load,
$w_u = 1.2 \times 1300 + 1.6 \times 1600 = 4120 \, \text{lb/ft}$
$d = 24 \, \text{in}$.
Concentrated load, $P = 5000 \, \text{lb}$
Factored, concentrated load, $P_u = 1.6 \times 5000 = 8000 \, \text{lb}$
Shear at mid-span, $V = 4000 \, \text{lb}$
Maximum shear at support, $V = \dfrac{1}{2}(4120 \times 30 + 8000) = 65{,}800 \, \text{lb}$

Shear force at critical location ($d$ from the support) is calculated using similar triangles:

$$\left(\frac{65{,}800 - V_u}{24''}\right) = \left(\frac{65{,}800 - 4000}{15' \times 12''}\right)$$
$$V_u = 57{,}560 \, \text{lb}$$

The shear force distribution for the factored load is shown in Figure 7.9. Using the simplified equation, shear carried by concrete:

$$V_c = 2\sqrt{f_c'}b_w d$$
$$V_c = 2\sqrt{4000} \times 12 \times 24 = 36{,}430 \, \text{lb}$$
$$\phi V_c = 0.75 \, V_c = 27{,}323 \, \text{lb}$$
$$\frac{V_u}{\phi} - V_c = \frac{57{,}560}{0.75} - 36{,}430 \, \text{lb}$$
$$= 40{,}317 < 8\sqrt{f_c'}b_w d. \text{ Therefore OK.}$$

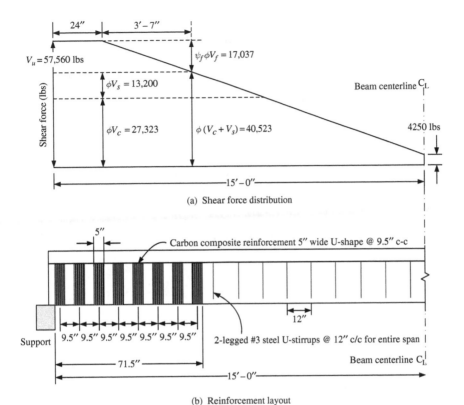

*Figure 7.9* (a) Shear force distribution and (b) shear reinforcement details for Example 7.1.

Shear taken by steel, $V_s = \dfrac{A_v f_y d}{s} = \dfrac{0.22 \times 40{,}000 \times 24}{12} = 17{,}600\,\text{lb}$

$\phi V_s = 13{,}200\,\text{lb}$

$\phi(V_c + V_s) = 40{,}523\,\text{lb}$

$V_u = 57{,}560 > \phi(V_c + V_s)$

Therefore, steel stirrups are *not* sufficient to carry the shear. Try U-shaped carbon composite for additional reinforcement. Shear to be carried by composite:

$$V_f = \frac{1}{\psi}\left[\frac{V_u}{\phi} - V_c - V_s\right] \text{ where } \psi = 0.85$$

$$V_f = \frac{1}{0.85}\left[\frac{57{,}560}{0.75} - 36{,}430 - 17{,}600\right]$$

$$V_f = 26{,}725\,\text{lb}$$

$$V_s + V_f = 44{,}325 < 8\sqrt{f'_c}b_w d. \text{ Therefore OK.}$$

Estimate effective fracture strain, $\varepsilon_{fe}$. First, effective depth for shear:

$$d_{fv} = d - h_f = 24 - 6 = 18\,\text{in.}$$

$$L_e = \frac{2500}{(n_f\, t_f\, E_f)^{0.58}}$$

Assume two layers with equivalent fiber thickness of 0.0065 in. per layer. $E_f = 33 \times 10^6$ psi and $\varepsilon_{fu} = 0.0167$. Therefore:

$$L_e = \frac{2500}{(2 \times 0.0065 \times 33 \times 10^6)^{0.58}} = 1.35\,\text{in.}$$

$$k_1 = \left(\frac{4000}{4000}\right)^{\frac{2}{3}} = 1$$

$$k_2 = \frac{d_f - L_e}{d_f} \text{ for U-wrap}$$

$$k_2 = \frac{18 - 1.35}{18} = 0.92$$

Therefore, reduction factor, $\kappa_V = \dfrac{k_1 k_2 L_e}{468\,\varepsilon_{fu}} \le 0.75$

$$\kappa_V = \frac{1 \times 0.92 \times 1.35}{468 \times 0.00167} = 1.589 > 0.75. \text{ Therefore } \kappa_V = 0.75$$

$$\varepsilon_{fe} = \kappa_V \varepsilon_{fu} = 0.0125 > 0.004. \text{ Therefore } \varepsilon_{fe} = 0.004$$

Try two-ply vertical strips that are 5-in. wide:

$$A_{fv} = 2 \times 0.0065 \times 2 \times 5 = 0.130\,\text{in.}^2$$

Spacing, $s_f = \dfrac{A_{fv} E_f \varepsilon_{fe}(1)d_{fv}}{V_f}$

$$s_f = \frac{0.130 \times 33 \times 10^6 \times 0.004 \times 18}{26{,}725} = 11.55\,\text{in.}$$

$$s_{f\,\max} = w_f + \frac{d_{fv}}{4} = 5 + \frac{18}{4} = 9.5\,\text{in.} < 11.55\,\text{in.}$$

Therefore $s_f = 9.5\,\text{in.}$

Since the strips are 5 in. wide, there will be a gap of 4.5 in. between strips.

Therefore proved 5-in. wide, U-strips of carbon composite at 9.5 in. center-to-center is adequate. This maximum reinforcement is needed only at the support. However, provide the same spacing for 6′0″ from the supports. Therefore, there will be eight strips that are 5 in. wide on each end. The steel reinforcement is sufficient for the center part of 18 ft. This arrangement is shown in Figure 7.9.

### Example 7.2: Strengthening for relocation of concentrated load

The beam shown in Figure 7.10 was designed to carry two point loads from mechanical equipment spaced 6 ft apart. The new equipment is of lower weight, and the location of point loads moved toward the center. The center 6 ft of the span does not contain any shear reinforcement because the shear capacity of concrete was sufficient to carry other loads. Design the strengthening system for the new location of concentrated loads. Assume a concrete strength of 3500 psi.

Solution:

The shear force distribution along the span is shown in Figure 7.11. The 6 ft in the mid-span region needs shear reinforcement. It is assumed that sufficient steel reinforcement is provided for the end sections.

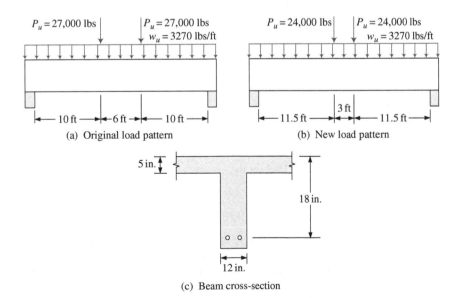

(a) Original load pattern

(b) New load pattern

(c) Beam cross-section

*Figure 7.10* Loading arrangement and cross-section for Example 7.2.

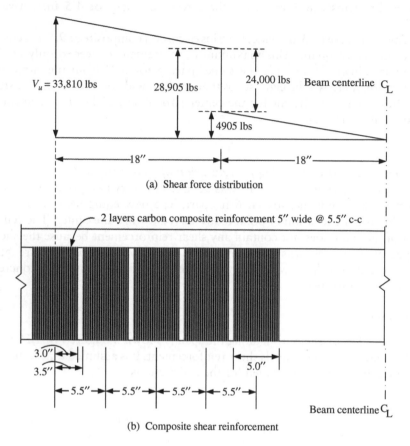

(a)  Shear force distribution

(b)  Composite shear reinforcement

*Figure 7.11* (a) Shear force distribution in midspan region and (b) shear reinforcement details for Example 7.2.

Shear capacity of concrete:

$$V_c = 2\sqrt{f_c'}b_w d = 2\sqrt{3500} \times 12 \times 18 = 25{,}558\,\text{lb}$$

$$\phi V_c = 0.75 \times 25{,}558 = 19{,}169\,\text{lb}$$

If shear force is less than $\dfrac{\phi V_c}{2} = 9584\,\text{lb}$, no reinforcement is needed. Maximum shear force of $33{,}810\,\text{lb} < \phi 8\sqrt{f_c'}b_w d = 76{,}674\,\text{lb}$. Therefore, composite reinforcement can be provided for 1'6" where increase in shear occurred due to movement of loads. Shear capacity needed from composite:

$$\psi V_f = \frac{V_u}{\phi} - V_c = \frac{33{,}810}{0.75} - 25{,}558 = 19{,}522\,\text{lb}$$

Assuming U-strips, $\psi = 0.85$. Therefore $V_f = \dfrac{19{,}522}{0.85} = 22{,}967\,\text{lb}$

Next, estimate effective fracture strain, $\varepsilon_{fe}$.

Effective depth for shear, $d_{fv} = d - h_f = 18 - 5 = 13\,\text{in.}$

$$L_e = \frac{2500}{(n_f t_f E_f)^{0.58}}$$

Assuming one layer of carbon composite with $E_f = 33 \times 10^6$ psi that is 0.0065 in. thick:

$$L_e = \frac{2500}{(1 \times 0.0065 \times 33 \times 10^6)^{0.58}} = 2.02\,\text{in.}$$

$$k_1 = \left(\frac{3500}{4000}\right)^{\frac{2}{3}} = 0.915$$

$$k_2 = \frac{d_{fv} - L_e}{d_{fv}} = \frac{13 - 2.02}{13} = 0.845$$

Reduction factor:

$$\kappa_V = \frac{k_1 k_2 L_e}{468 \varepsilon_{fu}}$$

$$\kappa_V = \frac{0.915 \times 0.845 \times 2.02}{468 \times 0.0167} = 0.20$$

$$\varepsilon_{fe} = 0.2 \times 0.0167 = 0.0033 < 0.004$$

Try vertical strips that are 5 in. wide:

$$A_{fv} = 2 \times 0.0065 \times 5 = 0.065\,\text{in.}^2$$

Spacing:

$$s_f = \frac{A_{fv} E_f \varepsilon_{fe} d_{fv}}{V_f}$$

$$s_f = \frac{0.065 \times 33 \times 10^6 \times 0.0033 \times 13}{22{,}967} = 4.01\,\text{in.}$$

Since the spacing is smaller than the width of each strip in Figure 7.12, more than one layer is needed to resist the shear. However, since the composite is spread out more evenly, to satisfy the shear requirements, the reinforcement should generate more than 22,967 lb over a distance (along the span) of $d_{fv}$, which is 13 in.

For 13-in. wide strip, shear capacity is:

$$2 \times 0.0065 \times 13 \times 0.0033 \times 33 \times 10^6 = 18,404\,\text{lb} < 22,967\,\text{lb}$$

Therefore a second layer of carbon sheet is needed. Try two layers that are 5 in. wide:

$$L_e = \frac{2500}{(2 \times 0.0065 \times 33 \times 10^6)^{0.58}} = 1.35$$

$$k_1 = \left(\frac{3500}{4000}\right)^{\frac{2}{3}} = 0.915$$

$$k_2 = \frac{13 - 1.35}{13} = 0.90$$

$$\kappa_V = \frac{0.915 \times 0.90 \times 1.35}{468 \times 0.0167} = 0.14$$

$$\varepsilon_{fe} = 0.14 \times 0.0167 = 0.0024 < 0.004$$

$$A_{fv} = 2 \times 2 \times 0.0065 \times 5 = 0.13\,\text{in.}^2$$

$$s_f = \frac{0.13 \times 33 \times 10^6 \times 0.0024 \times 13}{22,967} = 5.83\,\text{in.}$$

$$s_{f\,\text{max}} = w_f + \frac{d_{fv}}{4} = 5 + \frac{13}{4} = 8.25\,\text{in.} < 5.5\,\text{in.}$$

Therefore $s_f = 5.5\,\text{in.}$

Therefore, provide two layers that are 5 in. wide at 5.5 in. center-to-center. Alternatively, in the 18-in. region, provide five strips with a total width of 25 in.

## 7.10  Problems

**Problem 7.1:** For the beam cross-section shown in Figure 7.12, estimate the shear force contribution of carbon fiber composite and total shear capacity, $\phi V_n$. Compressive strength of concrete is 4000 psi. Assume U-wrap configuration, $\varepsilon_{fu} = 0.0167$, $E_f = 33 \times 10^6$ psi, and fiber thickness is 0.0065 in. (unidirectional). The steel stirrups consist of two-legged No. 3 bars at 12 in. center-to-center.

a)  One ply, continuous strip
b)  One ply, 5 in. wide at 6 in. center-to-center
c)  Two plies, 5 in. wide (each) at 12 in. center-to-center
d)  One ply, 5 in. wide at 12 in. center-to-center

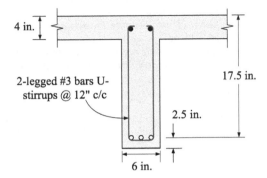

*Figure 7.12* Beam cross-section for Problem 7.10.

## References

ACI Committee 318 (2005) *Building Code Requirements for Structural Concrete and Commentary (ACI 318-05).* Farmington Hills, MI: American Concrete Institute, 430 pp.

ACI Committee 440 (2002) *Guide for the Design and Construction of Externally Bonded FRP Systems for Strengthening Concrete Structures (ACI 440.2R-02).* Detroit, MI: American Concrete Institute, 45 pp.

Anil, O. (2006) Improving shear capacity of RC T-beams using CFRP composites subjected to cyclic load. *Cement and Concrete Composites,* 28(7), pp. 638–649.

Chajes, M.J., Januska, T.F., Mertz, D.R., Thomson, T.A., and Finch, W.W. (1995) Shear strengthening of reinforced concrete beams using externally applied composite fabrics. *ACI Structural Journal,* 92(3), pp. 295–303.

Kachlakev, D. and McCurry, D.D. (2000) Behavior of full-scale reinforced concrete beams retrofitted for shear and flexural with FRP laminates. *Composites Part B: Engineering,* 31(6–7), pp. 445–452.

Khalifa, A., Gold, W., Nanni, A., and Abdel-Aziz, M.I. (1998) Contribution of externally bonded FRP to the shear capacity of RC flexural members. *ASCE Journal of Composite Construction,* 2(4), pp. 195–202.

Limbrunner, G.F. and Aghayere, A.O. (2007) *Reinforced Concrete Design.* 6th edn. Upper Saddle River, NJ: Pearson Prentice Hall, 521 pp.

MacGregor, J.G. and Wight, J.K. (2005) *Reinforced Concrete: Mechanics and Design.* 4th edn. Upper Saddle River, NJ: Pearson Prentice Hall, 1132 pp.

Malvar, L.J., Warren, G.E., and Inaba, C. (1995) Rehabilitation of Navy pier beams with composite sheets. In: *Proceedings of the Second FRP International Symposium, Nonmetallic (FRP) Reinforcements for Concrete Structures,* Ghent, Belgium: E&FN Spon, pp. 533–540.

Mosallam, A.S. and Banerjee, S. (2007) Shear enhancement of reinforced concrete beams strengthened with FRP composite laminates. *Composites Part B: Engineering,* 38(5–6), pp. 781–793.

Nawy, E.G. (2003) *Reinforced Concrete: A Fundamental Approach.* 5th edn. Upper Saddle River, NJ: Pearson Prentice Hall, 821 pp.

Norris, T., Saadatmanesh, H., and Ehsani, M.R. (1997) Shear and flexural strengthening of R/C beams with carbon fiber sheets. *Journal of Structural Engineering,* 123(7), pp. 903–911.

Priestley, M.J.N. and Seible, F. (1995) Design of seismic retrofit measures for concrete and masonry structures. *Construction and Building Materials*, 9(6), pp. 365–377.

Priestley, M.J.N., Seible, F., and Calvi, G.M. (1996) *Seismic Design and Retrofit of Bridges*. New York: Wiley-Interscience, 704 pp.

Seible, F. (1995) Repair and seismic retest of a full-scale reinforced masonry building. In: *Proceedings of the 6th International Conference on Structural Faults and Repair*, pp. 229–236.

Setareh, M. and Darvas, R. (2007). *Concrete Structures*. 1st edn. Upper Saddle River, NJ: Pearson Prentice Hall, 564 pp.

Triantafillou, T.C. (1998) Composites: a new possibility for the shear strengthening of concrete, masonry and wood. *Composites Science and Technology*, 58(8), pp. 1285–1295.

Wang, C., Salmon, C.G., and Pincheira, J.A. (2007) *Reinforced Concrete Design*. 7th edn. NJ: John Wiley & Sons, 948 pp.

# 8   Columns

## 8.1   Introduction

High-strength composites have been very effective in retrofitting columns to improve earthquake resistance. A large number of investigations conducted in the laboratory and field applications in buildings and bridges established the viability of composites for improving the performance of axially loaded members (ACI Committee 440, 2002). The composites are essentially used to confine the concrete, resulting in improvement of (i) compressive strength or axial force capacity, (ii) flexural and shear strength, (iii) flexural ductility, and (iv) performance of lap splices. Composites can also be used to repair and rehabilitate cap beams and beam–column joints.

A number of projects were carried out in the 1990s by the California Transportation Department on enhancing the earthquake resistance of columns (Policelli, 1995). The confinement was achieved by hand lay-up of jackets made from layers of epoxy-impregnated fiberglass fabrics or machine winding of epoxy-impregnated carbon fiber tows. Confinement was also achieved by placing prefabricated fiberglass shells around the columns and grouting the space between columns and shells. Typical installation techniques are shown in Figure 8.1. Columns and other axially loaded members in parking garages and multistory buildings are the other popular structural components chosen for repair and rehabilitation.

This chapter deals with the fundamentals of confined concrete in the behavior, analysis, and design of composite systems to enhance the performance of columns. The design principles were adopted from the procedure used for steel jacketing (Priestley *et al.*, 1996).

## 8.2   Behavior of confined concrete

It is well established that confined concrete can sustain much higher compressive strength and strain than unconfined concrete. The confinement prevents the expansion of uniaxially loaded specimens, contributing to higher load capacity. In the case of concrete confined by high-strength composites,

Figure 8.2, the confinement is passive. At low load levels, the expansion of concrete is proportional to Poisson's ratio, and the magnitude is small.

Therefore, the stresses induced in the composite are small until the longitudinal cracks develop in concrete. Depending on the type of concrete, the longitudinal cracks can occur at load levels ranging from $0.5f_c'$ for low-strength concrete to $0.9f_c'$ for very high-strength concrete. Typical behavior is shown in Figure 8.3.

The right side of Figure 8.3 represents the stress-uniaxial strain and the left side presents the lateral strain or expansion. After longitudinal cracks develop, the specimen expands rapidly. This phenomenon induces stresses in the confining reinforcement. The amount of confinement pressure will depend on the stiffness and thickness of the confining reinforcement. Increase in confinement pressure will result in increased compressive strength and strain, as shown in Figure 8.4. The challenge is to estimate the lateral pressure and the corresponding increase in strength and strain capacity.

*Figure 8.1* Retrofitting with composite material jackets: (a) hand lay-up of fiberglass with epoxy by the authors; (b) machine winding of carbon fiber and epoxy (Source: XXsys Technologies, Inc., 1997).

*Figure 8.1* (Continued).

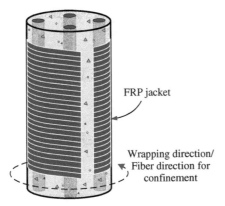

*Figure 8.2* Schematic of an FRP wrapped column showing fiber orientation (Source: Master Builders, Inc. and Structural Preservation Systems, 1998).

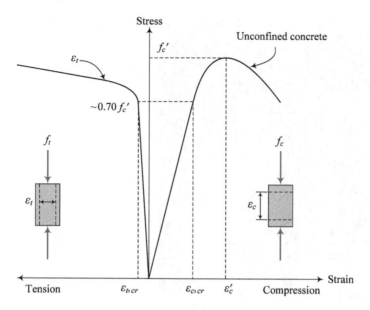

*Figure 8.3* Typical relationship for uniaxially loaded unconfined concrete showing stress versus longitudinal, transverse, and volumetric strain (Source: Master Builders, Inc. and Structural Preservation Systems, 1998).

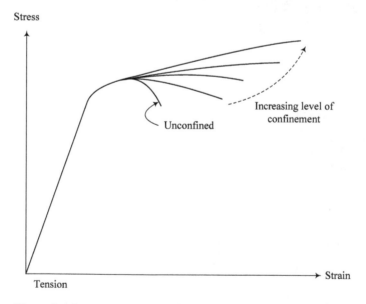

*Figure 8.4* Stress–strain curves for concrete under various levels of confinement (Source: Master Builders, Inc. and Structural Preservation Systems, 1998).

The behavior of confined concrete under lateral pressure is predicted using energy principles. If the confinement is provided by passive systems that depend on the expansion, the problem becomes more complex. The method proposed in this chapter is based on experimental results, and their interpretation is based on physics and regression analysis.

## 8.3 Behavior of columns confined with composites: concentric loads

### 8.3.1 Circular cross-sections

As discussed earlier, the cracking of concrete induces tension in the composite hoop. The first step is to establish equations for computing confining pressure. The free-body diagram of a circular cross-section confined by a composite whose thickness is $t_j$ is shown in Figure 8.5.

Using the principles of strength of materials, confining pressure to induce on the column, $f_{cp}$ can be expressed as a function of composite thickness and stiffness.

$$f_{cp} = \frac{2t_j f_f}{D} \qquad (8.1)$$

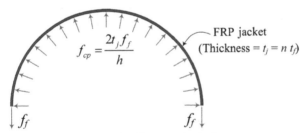

$$f_{cp} = \frac{2t_j f_f}{h}$$

FRP jacket
(Thickness = $t_j$ = $n\,t_f$)

$f_f$        $f_f$

(a) Free-body diagram of half composite jacket

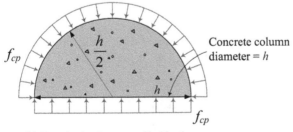

$f_{cp}$

$\dfrac{h}{2}$

$h$

Concrete column
diameter = $h$

$f_{cp}$

(b) Free-body diagram of half column

*Figure 8.5* Free-body diagram showing the internal and external forces on the FRP jacket and concrete column (Source: Master Builders, Inc. and Structural Preservation Systems, 1998).

where $D$ is the diameter of the column and $f_f$ is the force generated by the composite. If $E_f$ is the modulus of the fiber (composite), $t_j$ is the equivalent thickness of fibers, $\varepsilon_f$ is the hoop strain, and $n_f$ is the number of layers, then:

$$f_f = n_f t_f E_f \varepsilon_f \tag{8.2}$$

Therefore:

$$f_{cp} = \frac{2 n_f t_f E_f \varepsilon_f}{D} \tag{8.3}$$

If the confining reinforcement ratio, $\rho_f$, is Expressed as a function of area and diameter:

$$\rho_f = \frac{4 n_f t_f}{D} \tag{8.4}$$

and

$$f_{cp} = \frac{\rho_f E_f \varepsilon_f}{2} \tag{8.5}$$

When the expansion of concrete induces sufficient strain on the composite, the composite fractures leading to fracture of the concrete. For members subjected to combined shear and compression, the effective fracture strain of the composite at failure, $\varepsilon_{fe}$, should be limited to a maximum of 0.004 and $0.75\,\varepsilon_{fu}$:

$$\text{or } \varepsilon_{fe} = 0.004 \le 0.75\,\varepsilon_{fu} \tag{8.6}$$

Therefore, at failure, the confining pressure:

$$f_{cp} = \frac{E_f \varepsilon_{fe} \rho_f}{2} \tag{8.7}$$

Based on experimental results (ACI Committee 440, 2002), the compressive strength at this confining pressure, $f'_{cc}$, can be expressed as:

$$f'_{cc} = f'_c \left[ 2.25\sqrt{1 + 7.9\frac{f_{cp}}{f'_c}} - 2\frac{f_{cp}}{f'_c} - 1.25 \right] \tag{8.8}$$

The strain corresponding to this peak stress:

$$\varepsilon'_{cc} = \varepsilon'_c \left[ \frac{6 f'_{cc}}{f'_c} - 5 \right] \tag{8.9}$$

where $f'_c$ and $\varepsilon'_c$ are the peak compressive stress and the corresponding strain for the unconfined concrete, respectively. If $\varepsilon'_c$ is not given, it can be estimated using the equation:

$$\varepsilon'_c = 1.71 \frac{f'_c}{E_c} \tag{8.10}$$

The enhanced compressive strength, $f'_{cc}$, can be used for the computation of column axial force capacity. A reduction factor of $\psi_f$ is recommended for the concrete contribution. The recommended value for $\psi_f$ is 0.95. For nonprestressed members, the nominal capacity of a column can be computed using the equations recommended in ACI code (ACI Committee 440, 2002).

For a column with spiral reinforcement:

$$P_n = 0.85 \left[0.85 f'_{cc} \psi_f (A_g - A_{st}) + A_{st} f_y\right] \tag{8.11}$$

For a column with lateral ties:

$$P_n = 0.80 \left[0.85 f'_{cc} \psi_f (A_g - A_{st}) + A_{st} f_y\right] \tag{8.12}$$

$A_g$ is the cross-sectional area of concrete and $A_{st}$ is the area of longitudinal reinforcement.

### 8.3.2 Noncircular cross-section

Even though rectangular columns confined with composite reinforcement provide considerable increase in ductility and some increase in axial force capacity, it is recommended that increase in axial force capacity be neglected for design purposes. The complexity of rectangular hoop and ties contribution to confinement and confinement geometry led to the above recommendation. This recommendation is valid for most noncircular sections.

### 8.3.3 Computation of axial force capacity of confined circular column

*Example 8.1: Axial force capacity of confined circular column*
Compute the axial force capacity of a 16-in. diameter column with 10 #7 bars and spiral reinforcement with #3 bars. The confinement consists of three layers of (a) carbon and (b) glass fiber sheets. Assume a fiber thickness of 0.0065 in. for carbon and 0.026 in. for glass. The modulus for carbon

fibers is $33 \times 10^6$ psi and for glass fiber is $10 \times 10^6$ psi. The fracture strains are 0.0167 and 0.05 for carbon and glass, respectively. The compressive strength of concrete and yield strength of steel are 5000 psi and 60,000 psi, respectively.

Solution:

(a) Carbon fiber confinement:
   Given:

$$D = 16 \text{ in.}$$

$$A_{st} = 10 \times 0.6 = 6 \text{ in.}^2$$

$$n_f = 3$$

$$t_f = 0.0065 \text{ in.}$$

$$f_c' = 5000 \text{ psi}$$

$$f_y = 60,000 \text{ psi}$$

$$\varepsilon_{fu} = 0.0167$$

$$E_f = 33 \times 10^6 \text{ psi}$$

Confinement fiber reinforcement ratio:

$$\rho_f = \frac{4n_f t_f}{D}$$

$$= \frac{4 \times 3 \times 0.0065}{16}$$

$$= 0.0049$$

Effective fracture strain of confining fibers:

$$\varepsilon_{fe} = 0.004 \leq 0.7 \times 0.0167 = 0.0117$$

Confining pressure:

$$f_{cp} = \frac{E_f \varepsilon_{fe} \rho_f}{2}$$

$$= \frac{33 \times 10^6 \times 0.004 \times 0.0049}{2}$$

$$= 323.4 \text{ psi}$$

Compressive strength of confined concrete:

$$f'_{cc} = f'_c \left[ 2.25\sqrt{1 + 7.9\frac{f_{cp}}{f'_c}} - 2\frac{f_{cp}}{f'_c} - 1.25 \right]$$

$$= 5000 \left[ 2.25\sqrt{1 + \frac{7.9 \times 323.4}{5000}} - \frac{2 \times 323.4}{5000} - 1.25 \right]$$

$$= 6932\,\text{psi}$$

$$A_g = \frac{\pi(16)^2}{4}$$

$$= 200.96\,\text{in.}^2$$

Nominal axial capacity of the column:

$$P_n = 0.85\,[0.85f'_{cc}\psi_f(A_g - A_{st}) + A_{st}f_y]$$

$$= 0.85\,[0.85 \times 6932 \times 0.95\,(200.96 - 6.0) + 6.0 \times 60,000]$$

$$= 1,233,610\,\text{lb}$$

or $P_n = 1233\,\text{kips}$

(b) Glass fiber confinement:

$$\rho_f = \frac{4n_f t_f}{D}$$

$$= \frac{4 \times 3 \times 0.026}{16}$$

$$= 0.0195$$

$$\varepsilon_{fe} = 0.004 \le 0.7 \times 0.05 = 0.035$$

$$f_{cp} = \frac{10 \times 10^6 \times 0.004 \times 0.0195}{2}$$

$$= 390\,\text{psi}$$

$$f'_{cc} = 5000 \left[ 2.25\sqrt{1 + \frac{7.9 \times 390}{5000}} - \frac{2 \times 390}{5000} - 1.25 \right]$$

$$= 7272\,\text{psi}$$

Nominal capacity of the column:

$$P_n = 0.85\,[0.85 \times 7272 \times 0.95(200.96 - 6.0) + 6.0 \times 60,000]$$
$$= 1,279,107\,\text{lb}$$

or $P_n = 1279\,\text{kips}$

A flowchart for the computation of axial force capacity of confined circular columns is presented in Figure 8.6.

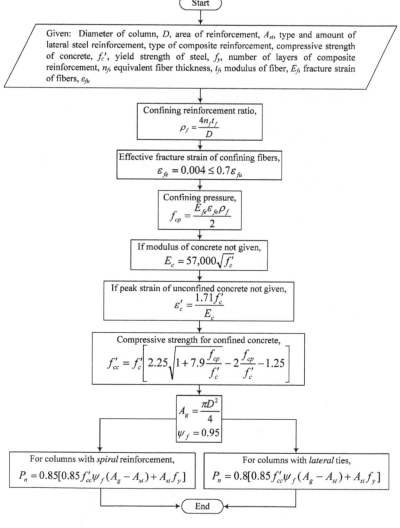

*Figure 8.6* Flowchart for computation of axial force capacity of confined circular columns.

## 8.4  Behavior of confined columns: eccentric loading

Confinement provided by FRP increases the strength and strain capacity of the concrete. When the column is loaded with concentric load, it can be safely assumed that FRP will provide a guaranteed increase in the strength capacity of the column. When the columns are subjected to eccentric load or a contribution of axial force and moment, the strains at the opposite sides of the column along the axis of the moment are not the same. The applied moment induces compression on one side and tension on the other side. As the magnitude of moment increases, the strain in the tension side keeps decreasing. In a typical reinforced concrete column, the increase in eccentricity is assumed to occur until the beam-column becomes a beam. In other words, the column could act as a beam loaded in the lateral direc-tion. In the case of columns confined with FRP, allowing excessive tensile strain may not be advisable because excessive cracking on the tension side might reduce the confinement effect. Note that fibers do not provide resist-ance along the axis of the column. Following a conservative approach, it is recommended to neglect the contribution of FRP, when the axial force in the column becomes zero. This aspect is further explained in the next paragraph.

As in the case of reinforced concrete, an interaction diagram relating the axial force and the corresponding moment capacity can be developed for the FRP confined column. A typical diagram is shown in Figure 8.7, with two lines representing a reinforced column with and without FRP confinement. The diagram for an FRP confined column is similar to the parent column with increased strength capacities. As explained earlier, the contribution of FRP is neglected when axial force is zero and, therefore, the moment capacities

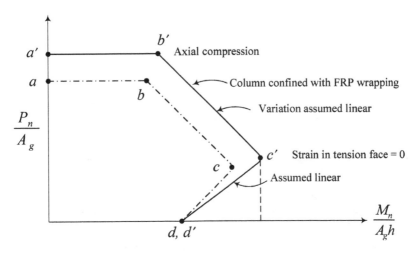

*Figure 8.7* Axial force-moment interaction diagram for columns confined with FRP.

for zero axial force are the same for both cases. In addition, the variation of axial force with respect to moment is assumed linear between the point of zero axial force and the maximum moment point (between points $c'$ and $d'$ in Figure 8.7). Between the points of $c'$ and $d'$, the contribution of FRP is gradually reduced. It is also recommended to limit the maximum allowable compressive strain of concrete to 0.005. This limit is necessary to maintain the shear integrity of the column.

### 8.4.1   *Construction of interaction diagram*

For most practical cases, the decrease of axial capacity with increase in eccentricity (or moment) can be assumed linear. Hence, an approximate interaction diagram can be constructed using the following guidelines.

*Step 1*: Estimate the maximum concentric axial force capacity. This value is obtained by multiplying the value $P_n$ obtained using Equation (8.11) or (8.12) by capacity reduction factor, $\phi$.

*Step 2*: Assume a linear strain variation starting with zero on one face of the column to the maximum strain at the other face of the column (Figure 8.8). The maximum strain should not exceed $\varepsilon'_{cc}$, computed using Equation (8.9), or 0.005.

Compute the force contribution of concrete and its location by integrating the curve over the circular section.

$$f_c = \frac{1.8 f'_{cc} \left( \dfrac{\varepsilon_c}{\varepsilon'_{cc}} \right)}{1 + \left( \dfrac{\varepsilon_c}{\varepsilon'_{cc}} \right)^2} \tag{8.13}$$

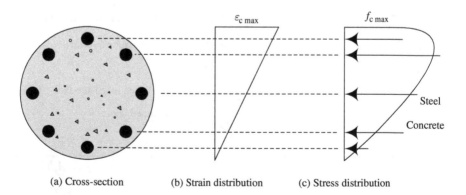

(a) Cross-section          (b) Strain distribution          (c) Stress distribution

*Figure 8.8* Strain and stress distribution for a typical beam column.

$\varepsilon_c$ and $f_c$ are the strain and the corresponding stress for concrete, respectively. The upper limit for the integral is the smaller value of $\varepsilon'_{cc}$ or 0.005. The location of the resultant force is computed by dividing the first moment of the force by the total force. It is recommended to perform numerical integration using a computer. If $C$ is the total compressive force and $\bar{y}$ is the location of the resultant from the neutral axis, then the axial force contribution of concrete is $C$ and the moment contribution is $C\bar{y}$.

*Step 3*: Using the linear strain distribution, Figure 8.8(b), compute the strain, stress, and force contribution of the reinforcing bars located across the cross-section. Addition of these forces to the concrete force contribution, $C$, provides the axial force capacity of the column. Similarly, multiply the force contribution of the bars with the lever arm to obtain the moment contribution. Summation of these moments and $C\bar{y}$ provides the moment capacity.

## 8.4.2 Approximate interaction diagram

Approximate interaction diagrams for compressive strength of 4000 psi are presented in Figures 8.9 and 8.10 for reinforcement location factors of

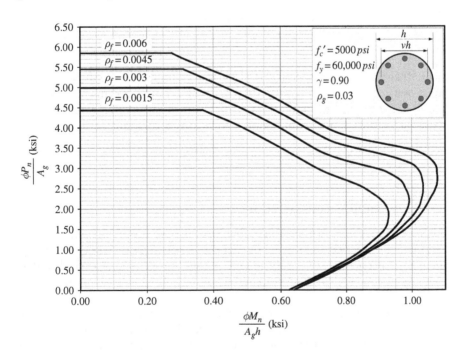

*Figure 8.9* Interaction diagram for a column with $f'_c = 5000$ psi, $\delta = 0.60$, and $\rho_g = 0.03$ wrapped with CF 130 FRP hoop reinforcement.

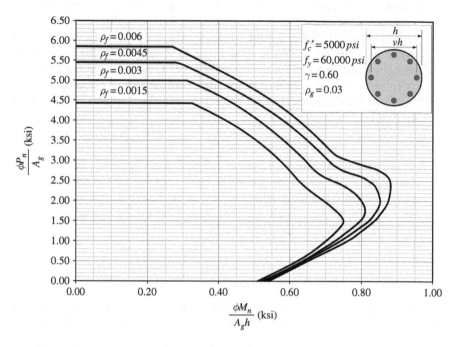

*Figure 8.10* Interaction diagram for a column with $f'_c = 5000$ psi, $\delta = 0.90$, and $\rho_g = 0.03$ wrapped with CF 130 FRP hoop reinforcement.

$\delta = 0.60$ and $\delta = 0.90$, respectively. These diagrams are valid only for column fibers with a modulus of $33 \times 10^6$ psi. Interaction diagrams for different compressive strengths, confinement reinforcement ratios, steel reinforcement ratios ($\rho_g = 0.03$), and reinforcement location factors can be found in the literature.

## 8.5  Serviceability considerations

When FRP jackets are used to increase the strength capacity of the columns, the service loads can be expected to increase correspondingly. Since FRP becomes active only after concrete develops cracks and undergoes expansion, FRP should be relied upon only for ultimate or occasional overloads.

The maximum stress in concrete under service loads should be limited to $0.65 f'_c$. It is assumed that concrete will not develop radial cracking up to this stress. The stress in steel at service loads should be limited to $0.60 f_y$ in order to prevent yielding of steel due to stress redistribution caused by creep of concrete under sustained and fatigue loads.

## 8.6  Design procedure for confined columns

As mentioned earlier, for concentric loads or loads with small eccentricity, the reinforced concrete column capacity can be safely increased by FRP confinement. In the case of columns subjected to combined axial force and moment, care should be taken to ascertain that cracking does not occur due to axial tension.

Columns may require strengthening or retrofit due to (i) change in load requirements, (ii) design deficiencies, (iii) construction errors, (iv) accidental physical damage, (v) long-term deterioration due to damage to concrete or corrosion of reinforcement, or (vi) a combination of these factors. In all cases, care should be taken to be sure that the existing column is properly repaired to ascertain the integrity of concrete, existing reinforcement, and the bond between these reinforcements and the concrete.

If the integrity of concrete is not 100%, the behavior may not provide the necessary expansion needed for the FRP. If excessive repairs are carried, the strength and stiffness of the repaired concrete should be determined. If the confinement is provided for a column with corroded reinforcement, the corrosion problem should be corrected first. This is very critical because the FRP jacket might hide the visual signs of further corrosion. The outside surface, on which FRP is placed, should be sound and smooth as per the manufacturers' specifications. If there is not proper contact, FRP will not provide the assumed confinement.

This section deals with the design of confinement only. It is assumed that the reader is familiar with the design of reinforced concrete columns. The design process can be carried out using the following steps:

*Step 1*: Using the information provided, compute the design moment, $M_u$, and axial load, $P_u$.
*Step 2*: Using the approximate interaction diagrams, estimate the confinement reinforcement ratio, $\rho_f$, as number of layers of confinement required.
*Step 3*: If necessary, check the capacities using the accurate method.
*Step 4*: Check the stresses for workload using the upgraded loads.

## 8.7  Design example

*Example 8.2: Design of the number of FRP layers for column confinement*

A circular column, 16 in. in diameter is reinforced with 10 – #7 bars and #3 spiral reinforcement with a clear cover of 1.5 in. Compressive strength of concrete and yield strength of steel are 5000 psi and 60,000 psi, respectively. The confinement is to be provided with carbon fibers with an equivalent thickness of 0.0065 in. and a modulus of $33 \times 10^6$ psi. The factored load and moment are 680,000 lb and 2,000,000 in.-lb, respectively.

Solution:

$$M_u = 2,000,000 \text{ in.-lb}$$
$$P_u = 680,000 \text{ lb}$$

The confinement reinforcement is estimated using the interaction diagrams, Figures 8.9 and 8.10.

Area of longitudinal steel, $A_{st} = 10 \times 0.6 = 6 \text{ in.}^2$

Area of concrete, $A_c = \frac{\pi}{4}(16)^2 = 201 \text{ in.}^2$

$$\rho_g = \frac{6}{201} = 0.03$$

Diameter of reinforcement centroid circle:

$$vh = 16 - 2 \times 1.5 - 2\left(\frac{3}{8}\right) - 2\left(\frac{7}{8}\right)$$
$$= 10.75 \text{ in.}$$

$$\delta = \frac{10.75}{16} = 0.67$$

$$\frac{P_u}{A_g} = \frac{680,000}{201} = 3400$$

$$\frac{M_u}{A_g h} = \frac{2,000,000}{201 \times 16} = 625$$

Using the interaction diagrams:

$$\delta = 0.6, \rho_g = 0.03, \rho_f = 0.003$$
$$\delta = 0.9, \rho_g = 0.03, \rho_f = 0.0015$$

Using linear interpolation for $\delta = 0.67$:

$$\rho_f \cong 0.026$$

Using an equivalent fiber thickness of 0.0065 in., the number of layers is:

$$n_f = \frac{0.0026 \times 16}{4 \times 0.0065} = 1.6$$

Therefore, use two layers of fiber sheets.

## 8.8   Rehabilitation of columns for earthquake resistance

Recent earthquakes in Asia, Europe, and the USA have demonstrated the vulnerability of columns in both buildings and bridges. Good design records kept by the California Transportation Department (USA) were very helpful to identify the design deficiencies. For example, prior to the 1971 San Fernando earthquakes, the transverse reinforcement in most columns consisted of #4 (0.5 in. diameter) spaced at 12 in. centers. It was found that this lack of transverse reinforcement was the main cause for most of the column failures. A careful study of the failed columns and the columns that successfully sustained the earthquake led to the following observations and conclusions.

- Lateral movement caused by the earthquakes created horizontal shear force resulting in excessive inclined cracking and a hinge at the bottom of the column. In addition, where the longitudinal reinforcement in the columns was not properly anchored, they were pulled out.
- Properly designed lateral reinforcement provides resistance to eliminate the above failure mechanisms. When the shear cracks cross the reinforcement, they not only resist the tension forces but also provide the needed ductility.
- When excessive spalling occurs at the bottom of the column due to movement, the lateral reinforcement reduces the buckling of the longitudinal bars and aids the formation of a plastic hinge. They also provide a clamping action for the lap splices, thus reducing the pulling out of longitudinal reinforcement.

Guidelines have been developed for properly designing the amount of lateral reinforcement needed and, more importantly, proper detailing. For example, spiral reinforcement or welded hoops were found to be much more effective than lateral ties. In the existing columns, the deficiencies can be corrected by proper rehabilitation. To design a proper rehabilitation scheme, an understanding of the failure mechanisms is essential.

## 8.9   Failure mechanisms for columns under earthquake loading

The most critical failure mode is the column shear failure, shown in Figure 8.11, which occurs due to excessive inclined cracking. The lateral movements cause excessive shear and inclining principal tensile stresses that are larger than the tensile strength of concrete.

If there is insufficient lateral reinforcement, these inclined cracks open rapidly and initiate cover spalling. Once the cover is lost, the transverse reinforcement could open or rupture, resulting in longitudinal reinforcement buckling and subsequent failure of the concrete core. Once excessive cracks occur in concrete, the subsequent events take place rapidly, resulting in very brittle or explosive failure. The remedial measure should focus on

(a)                                          (b)

*Figure 8.11*  Column shear failures, Northridge Earthquake 1994: (a) I-10 Santa
Monica Freeway; (b) I-118 Mission/Gothic undercrossing (Source:
Priestley, M. J. N., Seible, F. and Calvi, G. M., *Seismic Design and
Retrofit of Bridges*, © 1996 and owner. Reproduced with permission
of John Wiley & Sons, Inc.).

increasing the shear capacity, especially at the end regions, potential plastic
hinge regions, and center portions of columns between hinges.

The second failure mode can be classified as confinement failure,
Figure 8.12.

*Figure 8.12*  Flexural plastic hinge failures, Northridge earthquakes 1994
(Source: Priestley, M. J. N., Seible, F. and Calvi, G. M., *Seismic
Design and Retrofit of Bridges*, © 1996 and owner. Reproduced
with permission of John Wiley & Sons, Inc.).

This failure occurs due to excessive flexural cracking. The sequences are: excessive flexural cracking, crushing of concrete cover and spalling, and buckling of the reinforcement or crushing of concrete due to excessive compressive forces. This failure mode, which shortens the plastic hinge zone, typically occurs with some displacement and ductility, and is concentrated at shorter regions of the column. Given a choice, this failure mechanism is preferable to the shear failure because of its ductile nature and less complete destruction. The ductility of the hinge regions can be further improved by providing more confinement. The primary purpose of confining reinforcement is to delay concrete spalling and, thus, buckling of longitudinal reinforcement. In addition, confinement increases the concrete strength at failure.

The third failure mode occurs in columns with lap splices. These splices are provided for ease of construction, in which the bars coming out of the foundation is spliced to the main reinforcement of the columns. The failure mechanism will be the same as the second failure mechanism, provided that debonding of the reinforcement lap splice is prevented. When vertical microcracks that occur at the cover concrete reach a certain limit, debonding of splice is accelerated. Debonding and splitting can accelerate cracking of cover and subsequent spalling, Figure 8.13. Here again, confinement of

*Figure 8.13* Lap splice bond failure during Loma Prieta Earthquake, 1989 (Source: Priestley, M. J. N., Seible, F. and Calvi, G. M., *Seismic Design and Retrofit of Bridges*, © 1996 and owner. Reproduced with permission of John Wiley & Sons, Inc.).

the concrete wall improves the behavior of the column. The rehabilitation is necessary in situations where the lap splice lengths are not adequate or lateral reinforcement is insufficient.

When rehabilitation is done, care should be taken to avoid all the potential failures. Strengthening one part of the column could shift the failure mode and location to another region. It is always advisable to improve the performance of the entire length of the column.

## 8.10   Location of failure-critical regions and rehabilitation scheme

The locations for the three failure modes are shown in Figures 8.14 and 8.15 for single and double bending, respectively. The following are the salient points:

- Shear strengthening is needed for the entire length of the columns, except for small gaps at the footing or cap-beam. These gaps are needed to allow for the hinge rotation without added strength or stiffness. In the case of carbon or glass jackets, this gap could be as small as 1 in. or less depending on the thickness of the jacket.
- The length over which shear strengthening is needed is designated as $L_v$. This length is subdivided into $L_v^i$ and $L_v^o$ representing shear-critical regions inside and outside the hinge locations, respectively. Recommended magnitude for $L_v^i$ is 1.5 times the dimension of the column in the loading direction.
- Primary and secondary confinement regions are designated as $L_{c1}$ and $L_{c2}$, respectively. The secondary confinement region is needed to avoid the formation of a hinge at the end of the primary hinge height. If $D$ is the dimension of the column in the loading direction and $L$ is the effective height of the column:

$$L_{c1} \text{ and } L_{c2} \geq \left\{ \begin{array}{c} 0.5D \\ \dfrac{L}{8} \end{array} \right\} \tag{8.14}$$

The effective heights, $L$, for single and double bending are marked in Figures 8.14 and 8.15. Note that hinge lengths depend on both column size and shear space.
- The confinement for the lap splice length, $L_s$, should be greater than actual splice length values. This length, however, should exceed the primary hinge length, $L_{c1}$, by at least 50%.

The jacket made of high-strength fibers and matrix can be wrapped using automated or manual processes. In the automated process, the impregnated

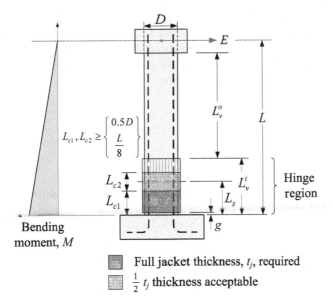

*Figure 8.14* Carbon jacket regions for bridge column retrofit, single-bending (Source: Priestley, M. J. N., Seible, F. and Calvi, G. M., *Seismic Design and Retrofit of Bridges*, © 1996 and owner. Reproduced with permission of John Wiley & Sons, Inc.).

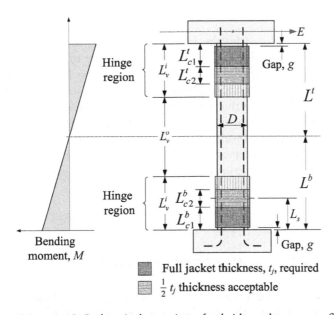

*Figure 8.15* Carbon jacket regions for bridge columns retrofit, double-bending (Source: Priestley, M. J. N., Seible, F. and Calvi, G. M., *Seismic Design and Retrofit of Bridges*, © 1996 and owner. Reproduced with permission of John Wiley & Sons, Inc.).

high-strength fiber is wrapped continuously around the entire column. In the manual process, prewetted fabrics or sheets are wrapped around the column. In both cases, the system should be designed to provide sufficient confinement. The amount of fibers needed depends on their modulus of elasticity and strain to failure. Design of composite systems for the rehabilitation of columns is discussed in the following sections.

## 8.11   Shear strengthening

### 8.11.1   *Shear resistance mechanism*

As in the case of beams, the shear strength contribution of columns can be divided into contributions from concrete, $V_c$, transverse reinforcement, $V_s$, and the horizontal component from the applied axial load compression strut between the column ends, $V_p$. The component $V_p$ adds to the shear strength of columns. Transverse reinforcement contribution could be from tie or spiral reinforcement. External FRP wrap can be used to increase the shear capacity. Even though the basic principles are the same for beams and columns, the shear resistance computation for columns is different from beams, primarily due to type of cross-sections. Cyclic loading conditions add to the complexity of computations. The overall nominal shear capacity:

$$V_n = V_c + V_s + V_p \tag{8.15}$$

The nominal capacity, reduced by the factor $\phi$ for shear, should be equal to or exceed external factored shear, $V_u$. The contribution of concrete, $V_c$, depends on the extent of damage or cracking. Within the hinge regions, $L_v$ (Figure 8.15), extensive damage can occur during a seismic event and, hence, the shear contribution in this region is much less than the shear contribution in the nonhinge region, $L_v^o$. The contribution in the hinge region will be a function of ductility level, $u_\Delta$. If the ductility level is high, more damage will occur. Using a conservative approach, the concrete shear contribution in hinge and nonhinge regions can be computed using the following equations:

$$\text{For nonhinge region, } V_c = 2\sqrt{f_c'} \times \text{effective shear area} \tag{8.16}$$

$$\text{For hinge region, } V_c = 0.5\sqrt{f_c'} \times \text{effective shear area} \tag{8.17}$$

Effective shear area can be taken as 80% of the gross column area. If the ductility level is low, the coefficient in Equation (8.17) can be increased to as high as 2.0. Some authors recommend a factor 3 instead of 2 for Equation (8.16). The contribution of lateral reinforcement can be computed using:

$$V_s = \frac{\pi A_h f_y D_c}{2s} \cot\theta \text{ (circular cross-section)} \qquad (8.18)$$

$$\text{or } V_s = \frac{n A_h f_y h_c}{s} \cot\theta \text{ (noncircular cross-section)} \qquad (8.19)$$

where $A_h$ is the area of one leg of horizontal reinforcement, $f_y$ is the yield strength of steel, $D_c$ is the core diameter of the circular column, $h_c$ is the core dimension in the loading direction, $s$ is the spacing of ties or the pitch of spirals, $n$ is the number legs of horizontal reinforcement in the loading direction, and $\theta$ is the angle between the principal compression strut to the column axis or the shear cracking inclination. Core dimension is measured from center to center of the peripheral horizontal reinforcement, as shown in Figure 8.16.

Principal strut inclinations can be assumed to be 30° and 45° for bridge columns and pier wall, respectively. Or, conservatively, $\theta$ can be assumed to be 45° for all cases, with $\cot\theta = 1$. The contribution of inclined compression strut:

$$V_p = P \tan\alpha \qquad (8.20)$$

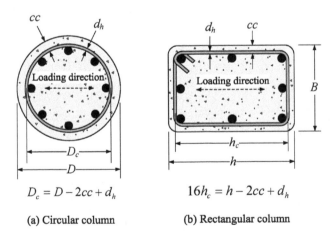

$$D_c = D - 2cc + d_h \qquad\qquad 16h_c = h - 2cc + d_h$$

(a) Circular column                    (b) Rectangular column

Figure 8.16 Definition of column core dimension. (Source: Priestley, M. J. N., Seible, F. and Calvi, G. M., Seismic Design and Retrofit of Bridges, © 1996 and owner. Reproduced with permission of John Wiley & Sons, Inc.).

where $P$ is the axial load on the column and $\alpha$ is the angle between the strut and the axis of the column. If $D$ and $L_c$ are the column dimension in the loading direction and the clear column height, respectively, then:

$$\tan \alpha = \frac{D - c}{2L_c} \text{ for single bending} \tag{8.21}$$

$$\tan \alpha = \frac{D - c}{L_c} \text{ for double bending} \tag{8.22}$$

where $c$ is the depth of the neutral axis measured from the extreme compression fiber, as shown in Figure 8.17.

### 8.11.2  Composite jackets for shear strengthening

If the factored total shear strength is less than the shear demand, $V_u^o$, then additional reinforcement is needed. If composites are used to provide this extra shear strength, the strength contribution:

$$\phi V_j \geq V_u^o - \phi[V_c + V_s + V_p] \tag{8.23}$$

where $\phi$ is the strength reduction factor (ACI Committee 318, 2005). Since the composite acts similar to reinforcement, its contribution can be estimated

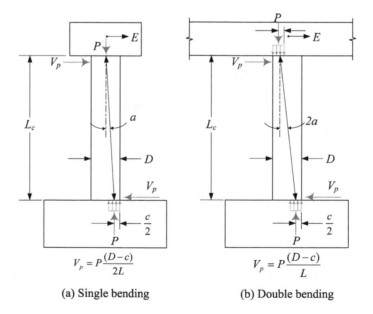

(a) Single bending          (b) Double bending

*Figure 8.17* Axial shear force contribution.

using equations similar to Equations (8.18) and (8.19). Since the fibers are wrapped around the existing column, the diameter of the column, $D$, is used instead of $D_c$. Similarly, for rectangular columns, $h$ is used instead of $h_c$. The force contribution of composite is a function of fiber area and fiber stress. Fiber stress, in turn, is a function of fiber modulus, $E_f$, and fiber strain, $\varepsilon$, when the maximum shear is present. Even though the fibers can sustain more than 1% strain, it is recommended to limit the usable strain to 0.4%. When the strains exceed 0.4%, the contribution of concrete was found to decrease rapidly. If $\cot \alpha$ is assumed as 1, and the equivalent fiber thickness in the jacket is $t_j$, then the contribution of composite jacket:

$$V_j = \frac{\pi}{2}t_j 0.004\, E_f D = 0.002\pi t_j E_f D \text{ for circular column} \tag{8.24}$$

$$V_j = 2t_j 0.004\, E_f h = 0.008\, t_j E_f h \text{ for noncircular column} \tag{8.25}$$

The equivalent fiber thickness of the jacket, $t_j$, can be obtained by multiplying the equivalent thickness of one layer by the number of layers. If tows are used, the area of tows over a unit height of column constitutes $t_j$.

The required fiber thickness, $t_j$, can be estimated using the following equations, obtained by combining Equations (8.23)–(8.25):

$$t_j = \frac{\frac{V_u^o}{\phi} - (V_c + V_s + V_p)}{0.002\pi + E_f D} = \frac{159}{E_f D}\left[\frac{V_u^o}{\phi} - V_c - V_s - V_p\right]$$

$$\text{for circular column} \tag{8.26}$$

$$t_j = \frac{\frac{V_u^o}{\phi} - (V_c + V_s + V_p)}{0.008 E_f h} = \frac{125}{E_f h}\left[\frac{V_u^o}{\phi} - V_c - V_s - V_p\right]$$

$$\text{for noncircular column} \tag{8.27}$$

Since the concrete contribution, $V_c$, depends on the location, two different jacket thickness need to be computed for plastic hinge region and nonplastic hinge region.

The shear demand depends on plastic column shear, or the shear at full overstrength, $V_u^o$. In the absence of reliable actual plastic shear information, $V_u^o$ can be conservatively estimated as 1.5 times the ideal shear capacity of the column at ductility, $u_\Delta = 1$. A flow chart for the computation of composite jacket thickness is presented in Figure 8.18.

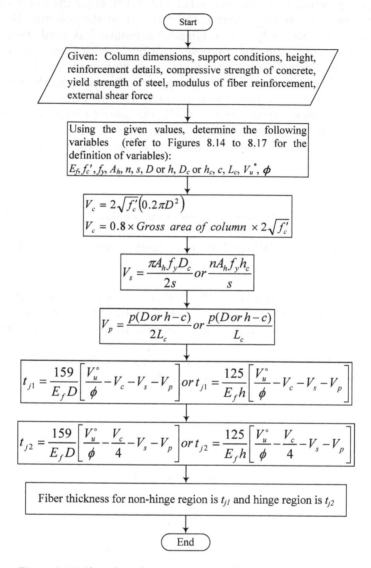

*Figure 8.18* Flowchart for computation of composite jacket thickness for shear strengthening.

### 8.11.3  Design example for computation of composite fiber thickness for shear strength

#### Example 8.3: Design of FRP thickness for shear

A circular column, 72 in. in diameter, is reinforced with 54 – #14 longitudinal bars and #4 ties at 12 in. centers. The yield strength of reinforcement and compressive strength of concrete are 60,000 and 6000 psi, respectively. Over-strength design shear based on maximum moment capacity is 2,600,000 lb.

The column, which is connected at both top and bottom to multicolumn bent and foundation, has clear height of 18 ft. Neutral axis depth at maximum response is 16.5 in. Axial load is 1,000,000 lb. Compute the number of carbon layers required to satisfy maximum shear requirements.

Solution:

Gross area of the column $= \dfrac{\pi}{4} (72)^2 = 4069 \text{ in.}^2$

For the hinge region, $V_c = 0.5\sqrt{f'_c} \times 0.8 \times$ gross area

$V_c = 0.5\sqrt{6000} \times 0.8 \times 4069 = 126,000 \text{ lb}$

For the nonhinge region, $V_c = 2\sqrt{6000} \times 0.8 \times 4069 = 504,000 \text{ lb}$

Assuming $\cot\theta = 1$, contribution of steel ties,

$$V_s = \frac{\pi A_n f_y D_c}{2s}$$

$A_h = 0.2 \text{ in.}^2$

$s = 23 \text{ in.}$

$f_y = 60,000 \text{ psi}$

Assuming a clear cover of 1.5 in.:

$D_c = 72 - 2(1.5 + 0.25) = 68.5 \text{ in.}$

$$V_s = \frac{\pi}{2} \times \frac{0.2 \times 60,000 \times 68.5}{12} = 107,560 \text{ lb}$$

As for strut action:

$$V_p = \frac{P(D - c)}{L_c}$$

$c = 16.5 \text{ in.}$

$L_c = 18 \text{ ft}$

Therefore:

$$V_p = \frac{1000,000 \ (72 - 16.5)}{18 \times 12} = 256,900 \text{ lb}$$

For a nonhinge region:

$V_c + V_s + V_p = 504,000 + 107,560 + 256,900 = 868,460 \text{ lb}$

For a hinge region:

$V_c + V_s + V_p = 126,000 + 107,560 + 256,900 = 490,460 \text{ lb}$

Assuming a carbon fiber modulus of $33 \times 10^6$ psi, and maximum strain of 0.004, thickness of carbon fibers required for nonhinge region:

$$t_j = \frac{159}{33 \times 10^6 \times 72} \left( \frac{2,600,000}{0.75} - 868,460 \right)$$

$$t_j = 0.174 \text{ in.}$$

Assuming an equivalent fiber thickness of 0.0065 in. per layer, number of layers required is:

$$n_f = \frac{0.174}{0.0065} = 26.7. \text{ Therefore, use 27 layers.}$$

For the hinge region:

$$t_j = \frac{159}{33 \times 10^6 \times 72} \left( \frac{2,600,000}{0.75} - 490,460 \right)$$

$$t_j = 0.199 \text{ in.}$$

$$n_f = \frac{0.199}{0.0065} = 30.6431$$

Number of layers $= 31$

Since the hinge region is $1.5D$, which is 9 ft, and the column height is 78 ft, the entire height of the column should be wrapped with 31 layers of 0.0065 in. thick carbon sheets.

## 8.12  Flexural plastic hinge confinement

During earthquake loading, hinges form at the maximum moment locations. In the case of columns and piers, these hinges form at foundation level and pier cap level in cases when the pier cap is connected to piers monolithically. These hinges should be able to sustain large inelastic rotations to avoid total and sudden failure. Confining the columns at critical locations is the most effective way to improve the rotation capacities. A number of studies have shown that carbon and glass fiber reinforced composites can be used for these types of confinements. The principles of design are the same as that of steel jackets. The major difference is the elastic behavior of high-strength fibers up to high strains and, hence, the better ability of hinges to sustain reversible loading. Investigations carried out by University of California – San Diego (Priestley et al., 1992; Seible et al., 1994a,b, 1995a,b,c,d) confirm that columns with carbon and fiberglass jackets can sustain reversed cyclic loading with very little knee formation. This aspect allows for better energy absorption (Figure 8.19).

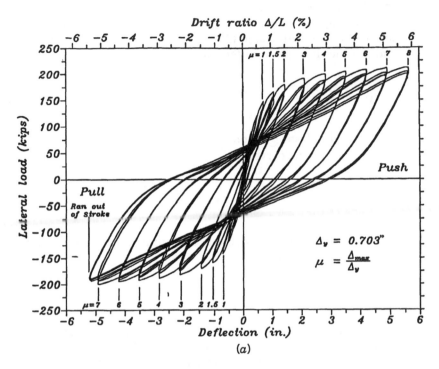

*Figure 8.19* Lateral force–displacement response of circular column-based lap splice retrofitted with a fiberglass-epoxy jacket (Source: Priestley *et al.*, 1996).

The primary guiding factor for jacket design is the confinement level that will provide the required maximum strain in concrete before failure. This required strain is a function of plastic rotation, $\varphi_p$, which, in turn, is a function of plastic curvature, $\phi_p$, and the length of the hinge, $L_p$. The following systematic procedure can be used for designing the jacket thickness.

*Step 1*: Establish the length of the plastic hinge. Since a gap, $g$, is left between the end of the jacket and the cap beams or foundations, the hinge will start forming in this gap. As more and more rotation occurs, the hinge that extends to the jacket is found to be a function of yield strength, $f_y$, and diameter of longitudinal bars, $d_{bl}$. Based on experimental and analytical investigations, the length of plastic hinge can be estimated using:

$$L_p = g + 0.3f_y d_{bl} \tag{8.28}$$

In Equation (8.28), $f_y$ and $d_{bl}$ are expressed in units of ksi and inches, respectively.

*Step* 2: Using plastic collapse analysis, establish the required plastic rotation capacity, $\varphi_p$. For earthquake loading, the requirement could be specified as the sustainable plastic displacement, $\Delta_p$. If $\Delta_p$ is specified and $L$ is the clear height of the column, then:

$$\varphi_p = \frac{\Delta_p}{L} \tag{8.29}$$

*Step* 3: Once the plastic rotation and the length of the hinge are established, the required plastic curvature:

$$\phi_p = \frac{\varphi_p}{L_p} \tag{8.30}$$

The total curvature, $\phi_m$, is obtained by adding $\phi_p$ to curvature at first yield of steel, $\phi_y$, to give:

$$\phi_m = \phi_p + \phi_y \tag{8.31}$$

The curvature at yield, $\phi_y$, depends on the cross-section and the reinforcement details. The reader is referred to textbooks on reinforced concrete for analysis (Limbrunner and Aghayere, 2007; MacGregor and Wight, 2003; Nawy, 2003; Setareh and Davras, 2007; Wang *et al.*, 2007). In most cases, the magnitude of $\phi_y$ will be two orders of magnitude smaller than $\phi_p$.

*Step* 4: Analyze the cross-section to determine the depth of the neutral axis, $c$, at ultimate load. Once $c$ is known, the maximum strain needed in the extreme concrete compression fiber, $\varepsilon_{cm}$, can be computed using:

$$\varepsilon_{cm} = c\phi_m \tag{8.32}$$

*Step* 5: The jacket should provide sufficient confinement to assure a maximum concrete strain of $\varepsilon_{cm}$. The FRP jackets made of carbon or glass should generate this maximum strain, $\varepsilon_{cm}$. The maximum strain that can be generated by FRP is found to be a function of maximum strain of unconfined concrete, confining stress generated by the jacket, and the compressive strength of confined concrete. If $f'_{cc}$ is the compressive strength of confined concrete, $\rho_f$ is the FRP ratio, and $f_{uj}$ and $\varepsilon_{uj}$ are the maximum stress and the strain for the jacket, respectively, the maximum strain can be written using equation:

$$\varepsilon_{cm} = 0.004 + \frac{2.5\rho_f f_{uj}\varepsilon_{uj}}{f'_{cc}} \tag{8.33}$$

For a circular column with diameter, $D$, the required jacket thickness, $t_j$ can be written as:

$$t_j = \frac{0.1(E_{cm} - 0.004)Df'_{cc}}{f_{uj}\varepsilon_{uj}} \qquad (8.34)$$

Based on tests conducted at the University of California at San Diego, confinement on rectangular columns was found to be at least 50% effective as compared to circular columns (Seible *et al.*, 1994a,b, 1995a). Based on the information, for rectangular columns, Equation (8.33) can be written as:

$$\varepsilon_{cm} = 0.004 + \frac{1.25\rho_f f_{uj}\varepsilon_{uj}}{f'_{cc}} \qquad (8.35)$$

The fiber reinforcement volumetric ratio for rectangular column:

$$\rho_f = 2t_j\left[\frac{b+h}{bh}\right] \qquad (8.36)$$

where $b$ and $h$ are width and thickness, respectively. Using Equations (8.34) and (8.35), for rectangular columns:

$$t_j = \frac{0.8(E_{cm} - 0.004)f'_{cc}}{f_{uj}\varepsilon_{uj}} \qquad (8.37)$$

Since the results are based on limited experiment results, it is recommended that the jackets be used only when the following conditions are satisfied:

$$P \leq 0.15f'_{cc}A_g \qquad (8.38)$$

$$\rho_l \leq 0.03 \qquad (8.39)$$

$$\frac{M}{Vh} \leq 3 \qquad (8.40)$$

The recommended maximum stresses and strains for carbon jackets are 150 ksi and 0.0125 in./in., respectively. The stress is based on composite thickness and not the equivalent fiber thickness. If carbon fiber thickness is used, the stress can be doubled. The designer should ascertain that these stresses and strains are achievable for the system chosen.

*Step* 6: Establish the extent of jacket for confinement. The extent of hinge height is shown in Figure 8.15. If $P/f'_{cc}A_g$ is greater than 0.3, increase the confinement length by 50%. The thickness of the jacket can be reduced by 50% for the second part of the hinge, marked $L_{c2}$ in Figure 8.15.

Alternative simplified approach:

An alternative approach is to design the jacket thickness based on a minimum confining pressure needed to generate a hinge. Using Equation (8.3), substituting $t_j$ for $n_f t_f$, and $\varepsilon_{fe}$ for $\varepsilon_f$, gives:

$$f_{cp} = \frac{2 t_j E_f \varepsilon_{fe}}{D} \tag{8.41}$$

or

$$t_j = \frac{f_{cp} D}{2 E_f \varepsilon_{fe}} \tag{8.42}$$

Note that Equation (8.42) is applicable only to circular columns, and $\varepsilon_{fe}$ is limited to $0.004 \le 0.75 \, \varepsilon_{fu}$. For most cases, the recommended confining pressure is 300 psi.

### 8.12.1   Design example for computation of composite thickness for hinge region

**Example 8.4: Design of FRP thickness for hinge region**

For Example 8.3, estimate the composite thickness using the following additional details. Maximum plastic displacement, $\Delta_p = 6$ in. Gap between jacket and connecting beams, $g = 2$ in. Concrete strength is 8000 psi. Use carbon fibers with a modulus of $33 \times 10^6$ psi, fiber strength of 300,000 psi, and solution fracture strain of 0.0125 in/in.

Plastic rotation, $\varphi_p = \dfrac{6.0}{18 \times 12} = 0.0278$

Length of hinge, $L_p = 2 + 0.3 \times 60 \times \dfrac{14}{8} = 33.5$ in.

Neglecting the curvature, $\phi_y$, $\phi_m = \phi_p$

$$= \frac{0.0278}{33.5} = 0.0008$$

Using the depth of neutral axis, $c = 16.5$ in.

Maximum concrete strain needed, $\varepsilon_{cm} = 0.0008 \times 16.5 - 0.0137$

$$t_j = \frac{0.1(0.0137 - 0.004)72 \times 8}{300 \times 0.0125} = 0.149 \text{ in.}$$

Using an equivalent fiber thickness of 0.0065 in., 23 layers are required. For practical reasons, carbon sheets with larger thickness should be used, so that a total thickness of 0.149 in. can be achieved using four or five layers.

Using a simplified alternative approach:

$$t_j = \frac{f_{cp}D}{2E_f\varepsilon_{fe}}$$

Assume $f_{cp} = 3000\,\text{psi}$,

$\varepsilon_{fe} = 0.004$

Therefore:

$$t_j = \frac{300 \times 72}{2 \times 33 \times 10^6 \times 0.004} = 0.082\,\text{in.}$$

For this thickness, 13 layers of 0.0065 in. thick carbon sheets are needed.

### 8.12.2  Check for buckling of longitudinal reinforcement

In order to prevent the buckling of bars, certain minimum confinement reinforcement is needed. The reinforcement ratio required, $\rho_f$, is a function of initial modulus of longitudinal steel, $E_i$, the secant modulus between yielding and ultimate load, $E_s$, modulus of confining reinforcement, $E_j$, and the number of longitudinal bars, $n_l$ (Priestley *et al.*, 1992).

$$\rho_f = \frac{0.45n_lf_s^2}{E_{ds}E_j} \tag{8.43}$$

where $f_s$ is the longitudinal steel stress at a strain of 0.04 and:

$$E_{ds} = \frac{0.4E_sE_i}{\left(\sqrt{E_{su}} + \sqrt{E_i}\right)^2} \tag{8.44}$$

For grade 60 steel, $f_s$ can be taken as 74,000 psi. For columns with grade 60 longitudinal bars and carbon fiber jackets, Equation (8.43) can be simplified as:

$$\rho_f = \frac{3.7n_l}{E_j} \tag{8.45}$$

or jacket thickness, $t_j \geq \dfrac{3.75n_lD}{4E_j}$ $\tag{8.46}$

The antibuckling check is needed for only slender columns, where $M/VD$ exceeds 4.

## 8.13  Clamping of lap splices

In some cases, longitudinal bars are spliced at foundation level for ease of construction (Figure 8.20). When overloading occurs, the longitudinal bars could pull out due to splice failure. The failure mechanism can be postulated as follows.

### 8.13.1  Failure mechanism for lap splices

The overloading creates two sets of cracks. The first set of cracks coagulate to form one continuous crack around the inside perimeter of the longitudinal bars (Figure 8.20). The second set of cracks that occur perpendicular to the first cracks are controlled by longitudinal reinforcement spacing. Typically, two cracks occur for every set of bars, one between the two spliced bars, and the second midway between the bars. When the uplift occurs, instead of bars being pulled out, the prisms between these cracks are pulled out. The resistance comes from the friction around the surface of the prisms.

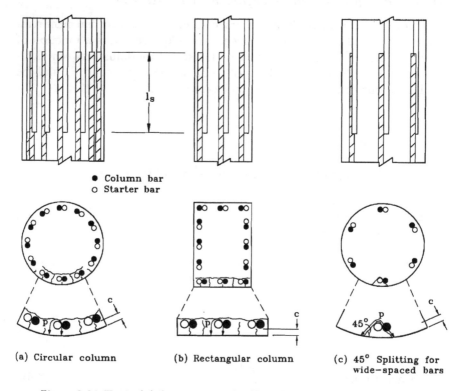

(a) Circular column     (b) Rectangular column     (c) 45° Splitting for wide-spaced bars

*Figure 8.20* Typical failure patterns for lap splice (Source: Priestley, M. J. N., Seible, F. and Calvi, G. M., *Seismic Design and Retrofit of Bridges*, © 1996 and owner. Reproduced with permission of John Wiley & Sons, Inc.).

If $n_l$ is the number of bars in the longitudinal direction, $d_c$ is the clear cover of longitudinal bars, the total length of crack surrounding one bar on the plan dimensions, $w_l$, is (Figure 8.20):

$$w_l = \left[\frac{p_r}{2n_l} + 2(d_c + d_b)\right] \tag{8.47}$$

where $d_b$ is the longitudinal bar diameter and $p_r$ is the inside perimeter along the longitudinal column reinforcement. If $L_s$ is the splice length, then the total concrete area that resists the pullout is:

$$\left[\frac{p_r}{2n_l} + 2(d_c + d_b)\right] L_s$$

The frictional force generated by this total crack area should be greater than the axial force capacity of the longitudinal bars. If the coefficient of friction for cracked concrete surface is assumed as 1.4 and the strain hardening of steel develops 40% more stress than yield stress, then confining stress needed to prevent pullout or lap splice failure, $f_\lambda$, can be obtained using:

$$f_\lambda = \frac{A_s f_y}{\left[\dfrac{p_r}{2n_l} + 2(d_c + d_b)\right] L_s} \tag{8.48}$$

### 8.13.2 Lap splice clamping design

To prevent lap splice failure, the confinement pressure provided at the bottom of the column should exceed $f_\lambda$ (Section 8.13.1). This confinement pressure could be from lateral reinforcement or a combination of lateral reinforcement and FRP jacket. If the lateral reinforcement ratio is low, its effect can be ignored. This is particularly true for columns with noncircular tie reinforcement. For columns reinforced with circular ties or spirals, the clamping pressure of $f_h$ can be estimated using:

$$f_h = \frac{0.002 A_h E_h}{D s} \tag{8.49}$$

where $A_h$ is the area of the hoop or spiral reinforcement, $E_h$ is the modulus of elasticity, $D$ is the column diameter, and $s$ is the hoop or spiral spacing. The diameter should be the spiral hoop inside diameter, but it can be closely approximated using the column diameter. To prevent any slip, it was found that the strains in the hoop reinforcement should be limited to 0.001. Hence, in Equation (8.49), the force contribution of hoop is computed using the strain 0.001 and the modulus of elasticity.

The contribution of the jacket, $f_j$, should exceed or be equal to $(f_\lambda - f_h)$. Again, assuming an effective strain of 0.001, if the modulus of the jacket material is $E_j$, and the thickness of the jacket is $t_j$:

$$f_j = \frac{0.002 t_j E_j}{D} \tag{8.50}$$

$$\text{or } t_j = \frac{f_j D}{0.002\,E_j} = \frac{500D}{E_j} f_j \tag{8.51}$$

$$t_j = \frac{500D}{E_j}(f_\lambda - f_h) \tag{8.52}$$

The jacketing method is not recommended for confining rectangular columns. However, if controlled debonding is permissible, twice the jacket thickness recommended for circular columns can be used. The aspect ratio for rectangular columns should be less than or equal to 1.5. The corners should be rounded to a minimum radius of 2 in. The jacket thickness should be computed using equivalent diameter, $D_e$. If $A$ and $B$ are the thickness and width of the columns ($A > B$), respectively:

$$D_e = \frac{b^2}{a} + \frac{a^2}{b} \tag{8.53}$$

where $a$ and $b$ are oval jacket dimensions, as shown in Figure 8.21.

$$a = kb \tag{8.54}$$

$$b = \sqrt{\left(\frac{A}{2k}\right)^2 + \left(\frac{B}{2}\right)^2} \tag{8.55}$$

$$\text{and } k = \left(\frac{A}{B}\right)^{\frac{2}{3}} \tag{8.56}$$

The lap splice retrofit should cover at least the splice length $L_s$.

### 8.13.3 Design example for lap splice clamping

*Example 8.5: Design of lap splice clamping*

A 36-in. diameter column is reinforced with 24 No. 11 bars in the longitudinal direction and No. 4 spiral with a spacing of 9 in. The longitudinal bars are spliced using a splice length of 60 in. Design the jacket thickness if carbon fibers with modulus of $33 \times 10^6$ psi are used. Assume a modulus of $29 \times 10^6$ psi and yield strength of 60,000 psi for both longitudinal and hoop steel. Clear cover for hoops is 2.5 in.

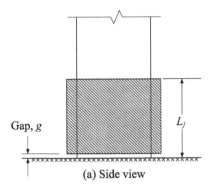

Gap, $g$        $L_j$

(a) Side view

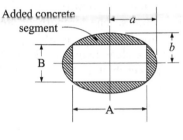

Added concrete segment

$a$

$b$

B

A

(b) Cross-section

*Figure 8.21* Confinement of rectangular column base with oval carbon jacket and added concrete segments.

Solution:

Area of longitudinal reinforcement, $A_s = 24 \times 1.56 = 37.44$ in.$^2$

Area of spiral reinforcement, $A_b = 0.2$ in.$^2$

Perimeter, $p_r = \pi\,[36 - 2(2.5 + 0.5 + 1.41)]$

$p_r = 92.7$ in.

$n_l = 24$

$d_b = 1.41$ in.

$d_c = 2.5 + 0.5 = 3$ in.

$$f_\lambda = \frac{37.44 \times 60{,}000}{\left[\dfrac{92.7}{2 \times 24} + 2(3 + 1.41)\right] 60} = 3482\,\text{psi}$$

$$f_b = \frac{0.002 \times 0.2 \times 29 \times 10^6}{24 \times 9} = 53.7\,\text{psi}$$

Composite thickness needed,

$$t_j = \frac{500 \times 36}{33 \times 10^6}(3482-53.7)$$

$$t_j = 1.87 \, \text{in}.$$

Since this thickness is very high, it may not be economical to provide carbon fiber jacket for this case.

## 8.14   Problems

**Problem 8.1:** A circular column, 60 in. in diameter is reinforced with 48 – #14 longitudinal bars and #4 ties at 9 in. centers. The strength of steel is 60,000 psi. Compressive strength of unconfined concrete is 5000 psi. The column is connected at top and bottom with a clear height of 20 ft. Axial load is 60,000 lb. Compute the carbon fiber jacket thickness to prevent shear failure. Assume a modulus of $33 \times 10^6$ psi for carbon fibers.

**Problem 8.2:** For problem 1, compute the jacket thickness for confining the hinge. Maximum plastic displacement $\Delta_p = 7$ in. Gap between jacket and support, $g = 2$ in. Strength of confined concrete is 7500 psi. Fiber strength equals 30,000 psi and fracture strain is 0.0125.

**Problem 8.3:** A 48-inch diameter column is reinforced with 18 – #11 bars in the longitudinal direction and #4 spirals at 6 in. centers. The longitudinal bars are spliced over a length of 50 in. Design the carbon jacket thickness using a modulus of $33 \times 10^6$ psi. Assume a modulus of $29 \times 10^6$ psi and yield strength of 60,000 psi for steel. This clear cover of hoop steel is 2.5 in.

## References

ACI Committee 318 (2005) *Building Code Requirements for Structural Concrete and Commentary (ACI 318-05)*. Farmington Hills, MI: American Concrete Institute, 430 pp.

ACI Committee 440 (2002) *Guide for the Design and Construction of Externally Bonded FRP Systems for Strengthening Concrete Structures (ACI 440.2R-02)*. Detroit, MI: American Concrete Institute, 45 pp.

Limbrunner, G.F. and Aghayere, A.O. (2007) *Reinforced Concrete Design*. 6th edn. Upper Saddle River, NJ: Pearson Prentice Hall, 521 pp.

MacGregor, J.G. and Wight, J.K. (2005) *Reinforced Concrete: Mechanics and Design*. 4th edn. Upper Saddle River, NJ: Pearson Prentice Hall, 1132 pp.

Master Builders, Inc. and Structural Preservation Systems. (1998) *MBrace Composite Strengthening System: Engineering Design Guidelines*. 2nd edn. Cleveland, OH: Master Builders, Inc.

Nawy, E.G. (2003) *Reinforced Concrete: A Fundamental Approach*, 5th edn. Upper Saddle River, NJ: Pearson Prentice Hall, 866 pp.

Policelli, F. (1995) *Carbon Fiber Jacket Wrapping of Five Columns on the Santa Monica Viaduct, Interstate 10 – Los Angeles*. Advanced Composites Technology Transfer/Bridge Infrastructure Renewal Consortium Report No. ACTT/BIR-95/14, August 1995, 110 pp.

Priestley, M.J.N, Seible, F., and Chai, Y.R. (1992) *Design Guidelines for Assessment Retrofit and Repair of Bridges for Seismic Performance*. Structural Systems Research Project, Report No. SSRP-92/01, University of California – San Diego, La Jolla, CA, August 1992, 266 pp.

Priestley, M.J.N., Seible, F., and Calvi, G.M. (1996) *Seismic Design and Retrofit of Bridges*. New York: Wiley-Interscience, 704 pp.

Seible, F., Regemier, G.A., Priestley, M.J.N. Innamorato, D., Weeks, J., and Policelli, F. (1994a) *Carbon Fiber Jacket Retrofit Test of Circular Shear Bridge Column, CRC-2*. Advanced Composites Technology Transfer Consortium Report No. ACTT-94/02, University of California – San Diego, La Jolla, CA, September 1994, 49 pp.

Seible, F., Regemier, G.A., Priestley, M.J.N., and Innamorato, D. (1994b) *Seismic Retrofitting of Squat Circular Bridge Piers with Carbon Fiber Jackets*. Advanced Composites Technology Transfer Consortium Report No. ACTT-94/04, University of California – San Diego, La Jolla, CA, November 1994, 55 pp.

Seible, F., Regemier, G.A., Priestley, M.J.N., Innamorato, D., and Ro, F. (1995a) *Carbon Fiber Jacket Retrofit Test of Rectangular Flexural Column with Lap-spliced Reinforcement*. Advanced Composites Technology Transfer Consortium Report No. ACTT-95/02, University of California – San Diego, La Jolla, CA, March 1995, 53 pp.

Seible, F., Regemier, G.A., Priestley, M.J.N., Ro, F., and Innamorato, D. (1995b) *Rectangular Carbon Jacket Retrofit of Flexural Column with 5% Continuous Reinforcement*. Advanced Composites Technology Transfer Consortium Report No. ACTT-95/03, University of California – San Diego, La Jolla, CA, April 1995, 52 pp.

Seible, F., Regemier, G.A., Priestley, M.J.N., Innamorato, D., and Ro, F. (1995c) *Carbon Fiber Jacket Retrofit Test of Circular Flexural Columns with Lap Spliced Reinforcement*. Advanced Composites Technology Transfer Consortium Report No. ACTT-95/04, University of California – San Diego, La Jolla, CA, June 1995, 78 pp.

Seible, F., Regemier, G.A., Priestley, M.J.N., and Innamorato, D. (1995d) *Rectangular Carbon Fiber Jacket Retrofit Test of a Shear Column with 2.5% Reinforcement*. Advanced Composites Technology Transfer Consortium Report No. ACTT-95/05, University of California – San Diego, La Jolla, CA, July 1995, 50 pp.

Setareh, M. and Darvas, R. (2007) *Concrete Structures*. 1st edn. Upper Saddle River, NJ: Pearson Prentice Hall, 564 pp.

Wang, C., Salmon, C.G., and Pincheira, J.A. (2007) *Reinforced Concrete Design*. 7th edn. New Jersey: John Wiley & Sons, 948 pp.

XXsys Technologies, Inc. (1997) *Automated Composite Column Wrapping: Process Overview Manual*. San Diego, CA: XXsys Technologies, Inc.

# 9 Load testing

## 9.1 Introduction

This chapter provides a method for efficiently and accurately assessing, in situ, the structural adequacy of reinforced and prestressed concrete building components. The guidelines allow the engineer to determine whether a specific portion of a structure has the necessary capacity to adequately resist a given loading condition. These guidelines establish a protocol for full-scale, in-situ load testing including planning, executing, and evaluating a testing program, which will assist the engineer in implementing an efficient load test.

### 9.1.1 General concepts and objectives

Valuable information regarding the health and performance of an existing structure may be gained by simply measuring its response to load. Traditionally, this view has been adopted in the implementation of load tests and structural monitoring. Both of these practices provide evaluations of structures that are much more representative than analytical approaches, especially when little is known about the structure's geometry and composition. However, it is difficult to justify the time and expense associated with full-scale load testing. To this end, rapid in-situ load testing takes the same approach to loading a structure and measuring its response, but the loads and measurements are specifically designed to reveal a certain characteristic of the structure. This approach allows for a much simpler evaluation that can be carried out in a fraction of the time and at a much lower cost.

Central to the concept of rapid load testing is the identification of the structural component and response that is of interest. For example, it may be of interest to investigate the bending capacity of a flat slab at mid-span of the column strip. The rapid load test would involve applying concentrated loads to the slab using hydraulic jacks. The location and magnitude of these loads are carefully chosen to produce critical responses in the structure while limiting the potential for causing permanent damage. The induced deflections and strains are measured, and the structure's performance is evaluated based on its response to loading.


## 9.1.2   *Background Information*

Load testing has long been a viable option for investigating a building, which exhibits "reason to question its safety for the intended occupancy or use" (Building Officials and Code Administrators International, 1987). Reports of full-scale in-situ load tests on buildings in the United States date back as far as 1910 (FitzSimons and Longinow, 1975). Committee 318 of the American Concrete Institute (ACI Committee 318, 1956) identified the need for guidelines for load testing of reinforced concrete (RC) structures for many years. Researchers (Bares and FitzSimons, 1975; Bungey, 1989; FitzSimons and Longinow, 1975; Fling, *et al.*, 1989; Genel, 1955a,b; Gold and Nanni, 1998; Hall and Tsai, 1989; Nanni and Gold, 1998a,b; RILEM Technical Committee 20-TBS, 1984) have investigated and discussed the methods of applying test loads and measuring structural response parameters. These investigations have attempted to refine the testing procedure over the years, but the fundamental protocol remains the same. Baseline measurements of the structural response parameters are taken before any loads are placed on the member. The structure is then loaded to a certain level and measurements are again recorded. Based on the measured response to loading, various acceptance criteria exist for determining the outcome of the test. Most existing criteria are only applicable to elements tested in flexure (ACI Committee 318, 2005; Bungey, 1989).

Currently, load testing protocols are included in various standards and specifications within the construction industry. The requirements of the ACI Committee 318 Building Code (2005) include provisions for static load testing of concrete structures. The general procedure required by ACI Committee 318 involves gradually applying the test load until a maximum load is reached and maintaining that load for 24 h. Measurements are recorded before any test load is applied, at the point of maximum load, after 24 h of constant loading, and 24 h subsequent to the removal of the test load. The structure is evaluated based on the maximum recorded deflection and the amount of deflection recovery.

In recent years, attempts have been made to change the way in which in-situ load tests are conducted. Some modifications to the load testing procedure defined in Chapter 20 of the ACI Committee 318 Building Code (1971) were suggested by a subcommittee formed within ACI Committee 437 (1991). One of those changes dealt with the duration of load application and time intervals between response measurements. It was proposed that the maximum test load need be applied for only 12 h with the final measurements taken 12 h after the load has been removed. However, there is still no scientific basis for either the 24- or the 12-h duration of load application. Long-term effects in concrete structures, such as creep, do not occur in a matter of hours or days. Rather, these effects only become significant after at least one month of constant load application (ACI Committee 318, 2005).

Recognizing that the long-term effects are not developed, the rapid load testing procedure shortens the duration of the load application to a matter of minutes. Loads are applied in quasi-static load cycles and the response of the structure is continually recorded. The rapid load testing procedure was originally developed to offer a nondestructive, yet conclusive, demonstration of the performance of new construction techniques and technologies. Among the first applications of the rapid load testing technique was the proof testing of externally bonded fiber reinforced polymer (FRP) sheets to strengthen concrete floor systems. In this application, much is known about the existing concrete structure. However, the relatively new FRP system is a cause for concern. Rapid load testing may be used to demonstrate the performance of the FRP systems in situ. The load test provides easy-to-understand physical results and an added degree of confidence.

This chapter deals with a protocol for establishing the duration of load application, the modality of the load cycles, and the criteria for rating the outcome of a rapid load test. The evaluation criteria are based on parameters that can be generalized for different types of load tests. In addition, such parameters can be computed during the execution of the test providing the means for real-time evaluation. Long-term effects that deal with the durability of concrete or other materials should be investigated using methods other than those proposed in this chapter.

## 9.2   Planning a rapid load test

A successful load test provides information essential to the assessment of the structural condition of the tested member. There are several phases involved in carrying out a successful load test. Information about the existing structure must first be gathered, the method of load testing must be determined, the load test must be carried out, and finally the results of the test must be analyzed and interpreted.

### 9.2.1   Evaluation of the structure

The preliminary steps in the planning of a load test are independent of the test type. These steps, clearly defined by ACI Committee 437 (1991) and the American Society of Civil Engineers (1991), include a study of drawings, reports, and calculations, verified by an on-site inspection, as well as determining the loading history and material characteristics of the structure.

### 9.2.1.1 Preliminary investigation

Before testing a given structure, a firm understanding of what is and is not known about the structure is required.

- *Structural geometry*: On-site inspections usually include the verification of dimensions and placement of reinforcement as shown on the as-built drawings, discussed in reports, and used in calculations by the designer.
- *Loading history*: The knowledge of the loading history may play a key role in assessing the structure's state. If damage has occurred, a study could reveal overloading or design and/or construction errors. If the use of a structure has changed, original and new load requirements must be determined.
- *Material characteristics*: Material properties, such as concrete compressive strength and reinforcing steel grade may have to be determined from samples collected in situ.

### 9.2.1.2 Definition of objectives

Rapid load tests can be run to determine the reserve capacity in a member that will undergo a change in usage or to quantify the level of damage in a member due to deterioration or other causes. They may also be used to establish the service levels of a unique design or to verify the functionality of novel materials. Questions about a member's capacity due to design or construction flaws may be answered through rapid load testing as well. The objectives of a rapid load test should address two issues:

- *Critical loading condition(s)*: The critical loading condition is that which is simulated by the rapid load test. For a meaningful evaluation of the structure, the critical loading condition is recommended to be at least 85% of its factored design loads minus the loads in place at the time of testing (e.g., self-weight). The critical loading condition for a given structure will depend on its intended use and design. As an example, using the load factors given by ACI Committee 318 (2005), the critical loading condition for a structure resisting uniformly distributed dead and live loads can be found from Equation 9.1,

$$w_{CT} = 0.85(1.2w_D + 1.6w_L) - w_{IP} \qquad (9.1)$$

where:

$w_{CT}$ = critical uniformly distributed loading condition to be simulated
$w_D$ = uniformly distributed dead load
$w_L$ = uniformly distributed live load
$w_{IP}$ = uniformly distributed load in place at time of testing

- *Critical response(s)*: The load test is meant to simulate selected responses induced by the critical loading condition. For example, the load test may induce the same bending moment at mid-span of a beam that the critical loading condition would induce. It is, therefore, necessary to determine what critical responses are of interest and need to be induced.

### 9.2.2   Test planning

Based on the objectives, a test plan should be developed that outlines the method of load application, the magnitude of the loads to be applied, and the measurements that will be taken during loading. This plan should reflect careful design and analysis performed by a qualified engineer. Guidance on the planning of these various aspects is given in the following sections.

#### 9.2.2.1   Selection of members

When considering a large population of members within a structure, the members tested should be representative of the structure in question. The most critical geometries for the most critical load cases should be represented in the selected member. For structures with repeated elements, it may be necessary to test a number of representative elements to arrive at a meaningful statistical conclusion as to the performance of the untested members. The engineer's judgment is critical to the proper selection of the type and number of elements to be tested.

Consideration must also be given to safety when selecting a test member. In certain instances, isolation of a member results in more predictable behavior of a system and ensures little damage to the rest of the structure. When the stiffness of adjacent elements adds significantly to the stiffness of the tested element, isolation may be very effective in lowering the maximum test load and reducing the chances of an alternate mode of structural failure.

#### 9.2.2.2   Methods of load application

The ideal load test would involve applying loads that exactly replicate the design load conditions. In this way, the resulting response of the structure is exactly as it would be under said loading. This is not always achievable, however. In the case of a floor or roof system, design loads are typically a uniform downward pressure. While this condition could be replicated by flooding the surface with water or stacking weights, such procedures are complicated and time consuming. More importantly, the load magnitude cannot be easily varied. Conversely, a rapid load test is based on the application of concentrated loads by means of hydraulic jacks. This method allows for rapid variation in the magnitude of the test load, which provides the means for cyclic loading of the structure. Consideration, however, should also be given to the fact that the hydraulic jacks used to supply the test load must be provided with an adequate reaction.

Depending on load magnitudes and the geometry of the structure, a suitable load application method should be selected. Examples of various types of load application methods (i.e., push-down, pull-down, closed loop, vehicle, and dropped weight) that allow the load magnitude to be easily varied are compared in Table 9.1 and are described below. The setup time (Table 9.1, column 2) includes installation of all loading devices and instrumentation used to monitor the structural behavior. These relative times do not take into account any work that may be involved in preparing the members for testing (e.g., removal of finishes, saw-cutting of members, drilling of holes, etc.), because such conditions are unique to a specific test and are independent of the test method. The minimum requirements (Table 9.1, column 3) are those that are essential to the loading of the test member. More equipment may be required depending on the conditions of the structure. The levels of on-site support (Table 9.1, column 4) include the activities that should be carried out by an on-site contractor. This includes operating forklifts, preparing test members, assisting in load setup, etc. The relative level of difficulty of load variation (Table 9.1, column 5) is based on the time it takes to change from one level of load to another. Another requirement for each test method is shown as the source of the reaction (Table 9.1, column 6). For each test method, the structural member used to provide the reaction must be carefully checked. These load application methods have limitations (Table 9.1, column 7) that should be considered.

- *Push-down test*: In the push-down test, one or more hydraulic jacks with extensions are used to provide the load that results in downward concentrated forces on the test member. Figure 9.1 shows an overall schematic of the push-down test method. In this figure, the shaded member is undergoing the load test in positive flexure.

  The extensions, attached to the hydraulic jacks will react against the ceiling when the jacks extend. Shoring is installed on one or more floors above the tested member to share the reaction. The displacement transducers (typically mounted on tripods) below the test member are used to measure the deflection at several points along the span of the member. Figure 9.2 shows a detail of the loading devices used in the push-down test method. Again, the shaded member is undergoing the load test. The plywood shown in this figure is used to protect the concrete from any localized damage. Also shown in Figure 9.2 are the hydraulic jack and the extensions used to apply the test loads. The extension cap is used as a centering device for the load cell.

  This test method has been used to verify the positive and negative flexural strengthening of prestressed and posttensioned flat slabs and to determine the shear capacity of RC ceiling joists. For additional information regarding members tested using the push-down method, see Gold and Nanni (1998), Nanni and Gold (1998a,b), and Nanni et al. (1998).

Table 9.1 Summary of load testing methods.

| Test method (1) | Setup time (2) | Minimum loading requirements (3) | On-Site support (4) | Load variation (5) | Source of reaction (6) | Limitations (7) |
|---|---|---|---|---|---|---|
| Push-down | Medium | • hydraulic jack and pump extensions to ceiling <br> • shoring of above floor(s) <br> • adequate weight of floors above for reaction | Little to none | Easy | Shored floor(s) above test member | Requires floor(s) above for reaction |
| Push-down (fixed reaction) | Medium | • hydraulic jack and pump hole in member <br> • high strength rod, chain, or cable <br> • high strength pulley <br> • adequate source of reaction | Little to none | Easy | Columns or piles below test member | Requires symmetric and close reaction points |
| Pull-down (mobile reaction) | Medium | • hydraulic jack and pump hole in member <br> • high strength rod, chain, or cable <br> • forklift or other mobile source of reaction | High to medium | Easy | Dead weight below test member | Maximum test load must be relatively low |

| | | | | | | |
|---|---|---|---|---|---|---|
| Closed loop | Long | • two hydraulic jacks and pump<br>• two holes in member<br>• high strength rod, chain, or cable source of reaction between test members<br>• adequately sized reaction beam | High | Easy | Internal structural member between tested members | Location and magnitude of load dependent upon size and length of reaction beam |
| Vehicle | Short | • forklift or other vehicle capable of carrying different amounts of load<br>• various amounts of weight | High to medium | Difficult | Not applicable | Load variation is time-consuming |
| Dropped weight | Short | • forklift or other device to carry load<br>• various amounts of weight | High to medium | Difficult | Not applicable | Load variation is time-consuming |

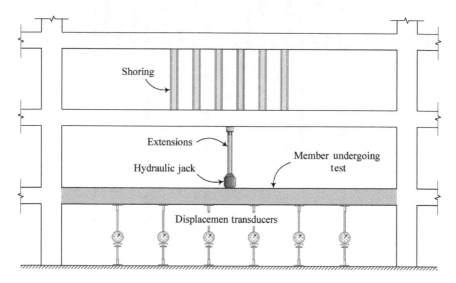

*Figure 9.1* Push-down test configuration.

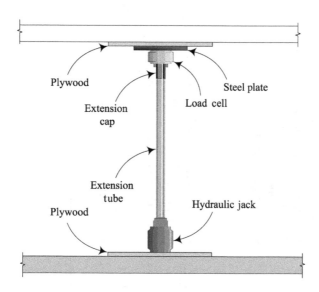

*Figure 9.2* Detail of push-down test method.

- *Pull-down test (fixed reaction)*: In the pull-down test with a fixed reaction, the reaction is provided below the tested member. Figure 9.3 shows a detail of the loading equipment in the pull-down test method. The darkly shaded member is undergoing a load test in positive flexure.

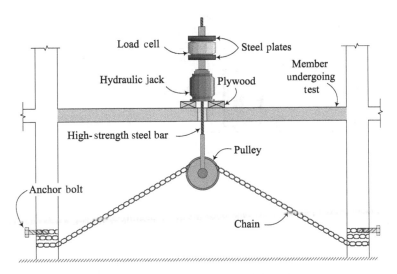

*Figure 9.3* Detail of pull-down test method with a fixed reaction.

The hydraulic jack shown in this figure applies the test load on the darkly shaded member with the reaction provided by the two columns. The displacement transducers, mounted on tripods on the floor below the test member, record the member's deflection at several locations. The high-strength steel bar is attached to the hydraulic jack and passed through the tested member.

At the end of the high-strength steel bar is a pulley, over which a chain passes. The chain is wrapped around columns on the floor below the tested member. Fire hoses were wrapped around the column in this case in order to protect them from any localized damage. As the hydraulic jack extends, the high-strength steel bar pulls up on the chain, which reacts against the columns, resulting in a downward concentrated load applied to the darkly shaded test member. Plywood is used to protect the concrete from any localized damage. The load cell measures the amount of load applied to the member during the test. This method has been used to verify the strengthening of an RC slab.

- *Pull-down test (mobile reaction):* In the pull-down test with a mobile reaction, the reaction is again provided below the tested member. Figure 9.4 shows an overall schematic of the pull-down test method using a mobile reaction. This figure is the view parallel to the span of the test member.

The darkly shaded roof member is undergoing a load test in positive flexure. Shown in this figure is the hydraulic jack, which provides the test load to the member by using the weight of the forklift as a reaction.

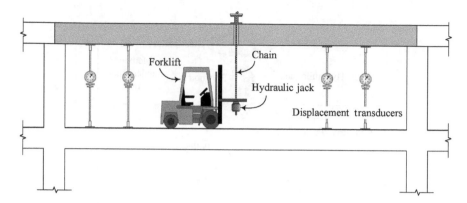

*Figure 9.4* Pull-down test configuration with a mobile reaction.

The hydraulic jack is connected to a chain, which passes through the test member and is attached to a spreader beam. The displacement transducers are mounted on tripods and record the deflection of the member at several points along the span of the test member. Figure 9.5 is a detail of the loading equipment used in the pull-down test method with a mobile reaction. This figure is the view perpendicular to the span of the test member.

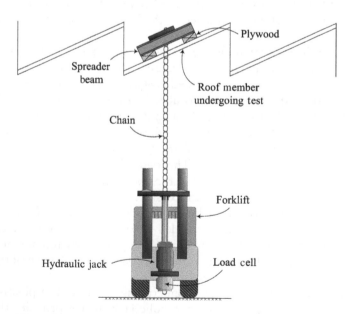

*Figure 9.5* Detail of pull-down test method with a mobile reaction.

The hydraulic jack, while extending, pulls down on the chain and subsequently the spreader beam. The spreader beam sits on plywood in order to protect the concrete from any localized damage. Again, the weight of the forklift is used as a reaction, and the load cell is used to measure the applied load. This method has been used to verify the strengthening of a prestressed concrete (PC) shell. For more information regarding members tested using the pull-down method with a mobile reaction, see Barboni *et al.* (1997), and Benedetti and Nanni (1998).

- *Closed loop*: The closed-loop test is the most elegant of all the choices because no external reaction is required. As shown in Figure 9.6, the closed-loop method is ideal for testing two members simultaneously if the locations of the loads are reasonably close. Shown in this figure are two hydraulic jacks, which apply the load to the two darkly shaded test members.

The inverted T-beam, located between the two test members, supplies the reaction. The displacement transducers measure the deflection at several points along the spans of both beams and they are located below the test members. Figure 9.7 shows a detail of the loading devices in the closed-loop method.

As the hydraulic jacks extend, they pull on the high-strength steel bars, which lift the reaction beam below the test members. Once the reaction beam comes into contact with the inverted tee beam, the resulting load is a downward force under each hydraulic jack. The plywood under

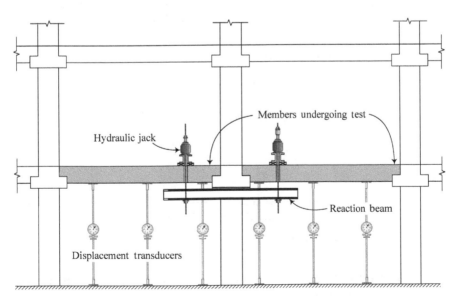

*Figure 9.6* Closed-loop test configuration.

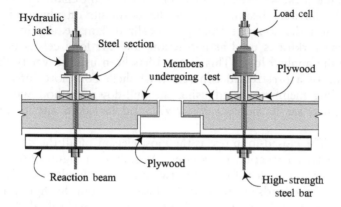

*Figure 9.7* Detail of loading devices for the closed-loop method.

the hydraulic jacks and between the inverted tee and the reaction beam are used to protect the concrete from any localized damage. The load cell measures the amount of load applied to one of the test members throughout the test. One requirement of this test method is that one pump must operate both hydraulic jacks in order to ensure equal loads at each jack. Figure 9.8 is a view of the closed-loop method perpendicular to the span of the beam in order to show how the concentrated load under each jack can be applied to the two stems of a double tee through the use of a steel section.

The steel section sits on plywood above the stem of the double tee in order to protect the concrete from any localized damage. This method

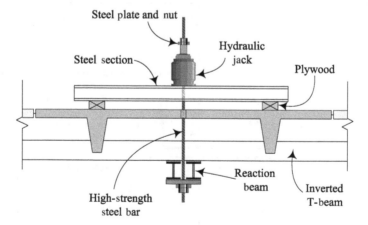

*Figure 9.8* The closed-loop method can be used to apply loads to both stems of a double tee.

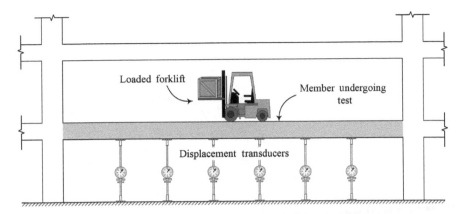

*Figure 9.9* Vehicle-loaded test configuration.

has been used to verify the strengthening of PC double-tee beams and PC joists in parking facilities. For more information about members tested using the closed-loop method, refer to Sawyer (1998), Hogue, *et al.* (1999a,b), Mettemeyer *et al.* (1999), Wuerthele (1999), and Gold *et al.* (2000).

- *Vehicle*: Figure 9.9 shows how a loaded forklift is used to apply load on the test member.

  The displacement transducers, located below the test member, record the deflection at several locations during the load test. The forklift, carrying a designated amount of weight, stops at several predetermined locations on the test member so that response measurements can be taken. Figure 9.10 shows a plan view of the forklift with the heavy axle centered on the test member.

  The displayed locations of the displacement transducers are those of a typical two-way system. This method has been used to verify the strengthening of PC decks in a power plant and a pier. For more information, see Bick (1998).

- *Dropped weight*: Figure 9.11 shows how a load test may be run by intentionally dropping a known amount of weight, from a given height, in a designated location and recording the structural response.

  The displacement transducers, which record the deflection of the member at several points, are located below the darkly shaded test member. Figure 9.12 shows a plan view of the forklift dropping the test load in the designated location. The positions of the displacement transducers may be those used if the test member is a two-way slab. The dropped weight test method has been used to verify the strengthening of a PC beam.

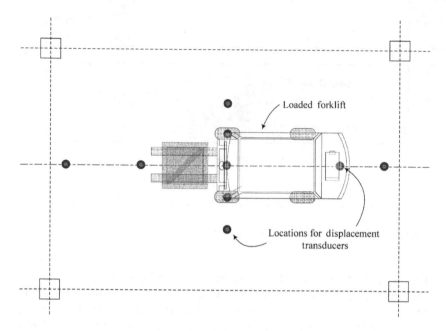

*Figure 9.10* Plan view of vehicle-loaded test method.

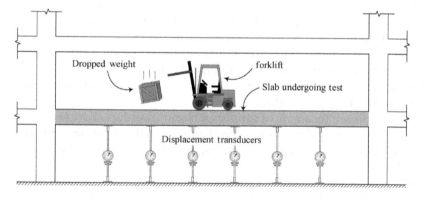

*Figure 9.11* Load test configuration for a dropped weight.

### 9.2.2.3 Test load magnitude

The magnitude of the test load is a function of the internal forces that need to be generated at the critical cross-section. As an example, a concentrated test load can be used to reproduce the same bending moments in a unit width of a slab as a uniformly distributed design load. The magnitude of the

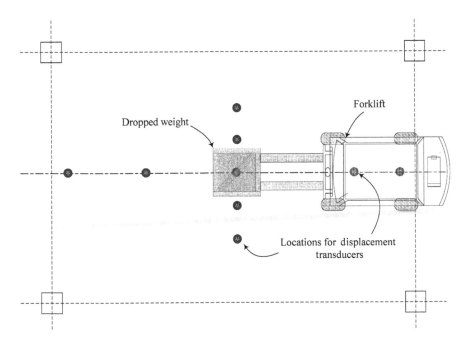

*Figure 9.12* Plan view of dropped weight test method.

concentrated load to accomplish this must be determined through careful structural analysis. The analysis to determine the test load magnitude should consider the following factors:

- *Load pattern*: The difference between the test load pattern and the pattern of the critical load condition should be considered. This is typically the difference between applying concentrated loads versus applying a uniformly distributed load.
- *Load sharing*: Often, test loads are not applied to adjoining elements that are loaded by the critical loading condition. These adjoining elements (structural or nonstructural) may contribute to the stiffness of the member being tested and share load with the tested member. Load sharing would include the stiffness of the elements in the orthogonal direction (e.g., two-way action in slabs). These effects must be considered in the analysis, which often precludes the use of simple one-dimensional structural models. Due to load-sharing effects, two- or three-dimensional modeling is typically required. Alternately, the member being tested may be physically isolated from adjacent elements. Figure 9.13 shows a double-tee beam that has been isolated from the adjoining tees by saw cutting along each joint.

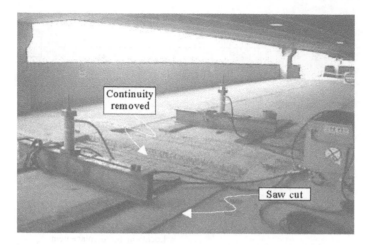

*Figure 9.13* Removing continuity allows for true simply supported condition.

By isolating the double tee in this fashion, the effects of load sharing are prevented and do not need to be considered in the analysis.

- *Boundary conditions*: The boundary conditions of the test member play an important role in determining the internal force distribution. Initial assumptions need to be made with regard to the degree of fixity of the supports. The degree of fixity can be correlated to a spring constant to be used in an analytical model (simple support, 0% fixity and a fixed end, 100% fixity). An estimation of the actual support fixity can be made after the structure has been loaded and the measurements of its deflection recorded.
- *Composite action*: Often, composite action (e.g., topping slabs) is not relied upon in determining responses induced by the critical load condition even though such action does exist in situ. The test load magnitude may need to be adjusted to, or the composite action be removed prior to the load test.
- *Temperature and environmental effects*: Variations in temperature and environmental conditions during testing may alter the monitored responses. Instruments need to be adjusted or the variations need to be avoided.

Once the magnitude and location of the maximum test load and the method of load application have been established, the strength of the element being tested with respect to other forces must be checked to ensure the safety. For example, if a test is meant to produce a critical flexural response, the shear capacity of the structure should be checked to prevent shear failure. If members within the structure are used to supply the reaction to the test load, the capacity of those members should be checked as well. It is important to recognize that any structural analysis must treat fixity and stiffness as assumptions. The preliminary structural analysis is used to estimate the magnitude of the test load only. Once the structure is loaded and its response is measured, the assumptions made in the analysis can be refined, based on the structure's actual behavior. With these refinements, the actual induced internal forces can be determined with a much higher degree of accuracy.

### 9.2.2.4 Prediction of structural responses

The preliminary analysis of the structure should also be used to predict the deformations of the structure to be measured in the field. Deformations under various load levels should be predicted to compare the deformations at various stages of loading during the load test. These predicted deformations allow the individual performing the test to determine whether the member is behaving as expected. Predicted deformations of concrete members are only used as approximations since the changing moment of inertia along the length of the test member makes exact calculations difficult. However, if the actual deformations are substantially different from the predicted deformations, then the load test can be stopped before any damage is done to the member. Further analysis may need to be performed to resolve the discrepancies.

### 9.2.3 Equipment

Advances in technology, in the area of load application and structural response measurements, have provided the means for this rapid load testing procedure. The following sections discuss some of the equipment that has been used to perform rapid load tests.

### 9.2.3.1 Hydraulic jacks and pumps

The equipment that has made rapid load testing possible is that which supplies the test load. Hydraulic jacks are used in rapid load testing because they are easy to install and control. Hydraulic jacks allow for relatively rapid variations in load, but the greatest advantage they have over using dead weights (e.g., water, bricks, etc.) is that the load applied by the hydraulic jacks can be removed instantaneously if a problem should arise during a load test. Figure 9.14 shows two 60,000-lb (267 kN) hydraulic jacks.

These jacks, including the steel bases, only weigh about 35 lb (156 N). Hydraulic jacks with a variety of capacities are available. The pump, which

*Figure 9.14* Hydraulic jacks used to apply load.

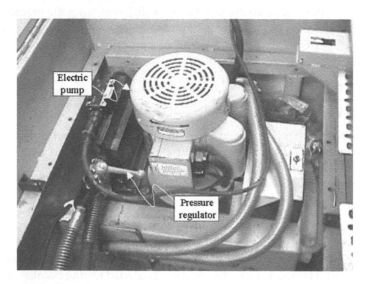

*Figure 9.15* Electrical pump used to supply fluid to hydraulic jacks.

supplies the fluid to the hydraulic jacks, can either be electrical, as shown in Figure 9.15, or manual, as shown in Figure 9.16. For an electric pump, the pressure regulator is used to control the amount of hydraulic fluid flowing to the jacks, which provides a controlled means of load application.

*Figure 9.16* Manual pump used to supply fluid to hydraulic jacks.

An electric pump usually has a control switch, which can be used for immediate removal of the applied load. For convenience, the electrical pump may be stored in a transport box as shown in Figure 9.17. The transport box, supported on wheels, may be used to move the hydraulic jacks and hoses, in addition to the pump, to the test site.

*Figure 9.17* Loading equipment.

### 9.2.3.2   *Instrumentation*

Some of the instrumentation used to monitor the behavior of a test member is summarized in Table 9.2. This table includes the common names for the devices, some of their suggested uses, recommended minimum measurable values, and measuring ranges. Additional information on these devices as well as many others is available in the literature (Bungey, 1989; Carr, 1993; Fraden, 1993).

Deflections are measured using linear variable differential transducers (LVDT), as shown in Figure 9.18. LVDTs are available in a variety of ranges and accuracy levels.

In order to reach test members, LVDTs are often mounted on tripods and/or placed on scaffolding. LVDTs used in rapid load testing should have a spring-loaded inner core, which allows the measuring head to return to some reference position.

*Table 9.2* Summary of instrumentation used in rapid load testing.

| Parameter | Devices | Recommended minimum measurable value | Measuring range |
|---|---|---|---|
| Deflection | LVDT | 0.0001 in. | ±2 in. |
| Rotation | Inclinometer | 0.001 deg | ±3 deg |
| Strain | Strain gage | 1 με | ±3000 με |
|  | extensometer | 50 με | ±10,000 με |
|  | LVDT | 50 με | ±10,000 με |
| Crack width | Extensometer | 0.0001 in. | ±0.2 in. |
| Load | Load cell | 10 lb | 0 • 200,000 lb |
|  | transducer | 100 lb | 0 • 200,000 lb |

Note: 1 in. = 25.4 mm, 1 lb = 4.445 N.

*Figure 9.18* LVDTs used to measure deflection (Source: RDP Electrosense, 2008).

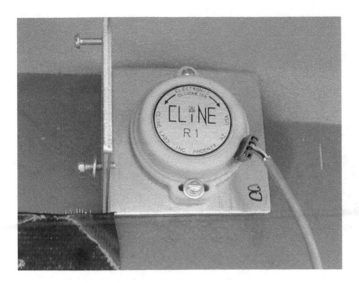

*Figure 9.19* Inclinometer used to measure slope (Source: Copyright © (2008a) California Department of Transportation).

Inclinometers, as shown in Figure 9.19, are used to measure the rotation or slope of a test member. Because values of slope can easily be correlated to deflections, these instruments can be essential when testing a member that is too tall for LVDT stands to reach. The inclinometer shown in this figure can be mounted on a variety of vertical and horizontal surfaces.

Strains in a test member can be measured in a variety of ways, depending on the level of accuracy and the expected magnitude. The most common method for measuring strain is via electrical resistance strain gages, which are bonded directly to the surface of the material for which the strain will be measured. Figure 9.20 shows an electrical resistance strain gage reading the compressive strain in concrete at the location of load application.

Strain gages can also be mounted on steel or other materials, which are expected to undergo tensile forces. Because they are susceptible to variations in temperature, most electrical resistance strain gages have temperature compensation coefficients. Electrical resistance strain gages are ineffective when they are intersected by a crack. To measure the strain over a crack, extensometers or LVDTs can be used. An extensometer, as shown in Figure 9.21, is attached directly to the surface on two knife-edges, which straddle an anticipated or existing crack.

An extensometer can then be used to either measure the average strain over the gage length between the two knife-edges or to measure the change in width of an intersecting crack. LVDTs can be used to determine the average strain over a larger gage length than that provided by the extensometer. Figure 9.22 shows how an LVDT can be used to measure strain. The horizontal LVDT is placed into a bracket, which is fixed to the test member.

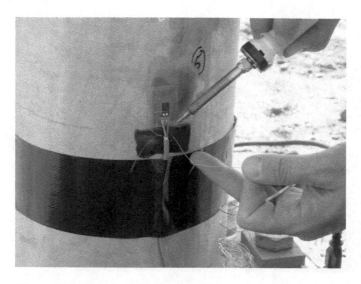

*Figure 9.20* Typical strain gage used to measure compression strain in concrete (Source: Copyright © (2008b) California Department of Transportation).

*Figure 9.21* Extensometers used to measure the strain across crack widths.

Another bracket is fixed to the test member such that the apparatus spans an existing or an anticipated location of a crack. The distance between the two brackets is the gage length over which the average strain is computed.

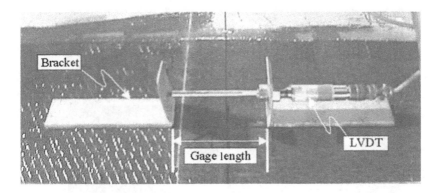

*Figure 9.22* LVDT used to measure the average strain over a gage length.

A device used to monitor the level of load application is a load cell, shown in Figure 9.23. As a requirement of many of the load testing methods described earlier, this load cell is donut shaped so that a high-strength steel bar may be passed through it.

Load cells come in a variety of shapes, sizes, and capacities. Another advantage to the use of hydraulic jacks is an additional means of monitoring the load they apply. Pressure transducers can be used to measure fluid pressures in the hydraulic system, which can be calibrated to a specific level of load.

*Figure 9.23* Load cells used to measure applied load.

*Figure 9.24* Data acquisition system collects data continuously from several
devices.

A data acquisition system, as shown in Figure 9.24, which is capable of collecting readings from several devices simultaneously as load is being applied,
is essential to the performance of a successful rapid load test.

The data acquisition system shown in this figure is capable of simultaneously collecting data from 24 separate devices, including pressure
transducers, load cells, LVDTs, inclinometers, extensometers, and strain
gages. The data acquisition unit is connected to a laptop computer, in which
all the collected data is stored. The data acquisition system, much like the
electric pump, is stored in an easy-to-handle transport box, which is taken
directly to the test site. The transport box, mounted on wheels for ease in
handling, is also equipped with additional storage compartments, in which
the measuring devices can be stored.

### 9.2.4 *Execution*

Not until a sufficient amount of planning has been done can the rapid load
test proceed. All parties involved should be aware of the levels of load to
which the member will be tested, and there should be a clear understanding
of the events that could occur that would lead to the termination of a load
test. All information regarding the load test should be clearly defined in a
"plan of action" that is distributed, in advance, to all the concerned parties.
The client or a representative should review the "plan of action" carefully
and give approval before the load test begins.

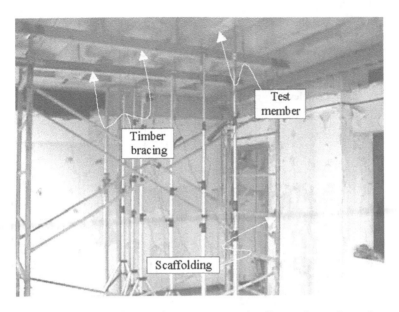

*Figure 9.25* Scaffolding used to prevent total collapse of tested member.

A concern in running a load test is the safety of the structure and those persons performing the load test. The use of scaffolding, shoring, straps, or chains may be key in the prevention of collapse of the member if a premature failure should occur. Figure 9.25 shows how scaffolding and timber bracing can be used for emergency support of the dead weight of the test member.

A safety measure should in no way interfere with the results of the load test. Only those persons essential to the load testing procedure should be in the area during the load test. Those persons performing the load test should always remain a safe distance from the test member. No individual should walk on or under a member being tested when test loads are being applied. Should it be essential to walk under or on a test specimen, the test load should first be decreased to a load that is deemed safe by the engineer in charge of the load test. Safety is essential to the performance of a successful load test.

## 9.3   Performance of a rapid load test

The protocol defined in the following sections is a generic guideline. Modifications may be made to meet the needs of a specific project.

### 9.3.1   *Procedure for a rapid load testing*

The rapid load test involves applying a load in quasi-static load cycles. The full test of a member is comprised of six load cycles with the modality shown

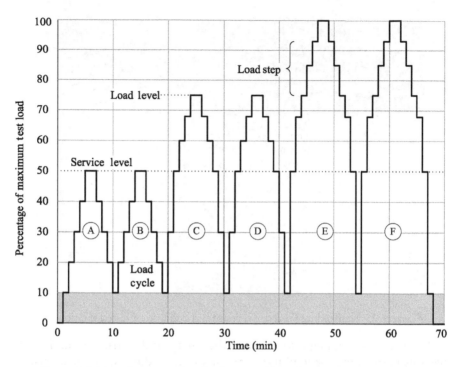

*Figure 9.26* Load steps and cycles for a rapid load test.

in Figure 9.26. In this figure, the vertical axis reports the applied test load as a percentage of the maximum value to be applied.

The horizontal axis reports the cumulative time, in minutes, as the test proceeds. Each individual load cycle (defined on Figure 9.26 by a circled letter) includes four to six load steps. Any given load cycle consists of an initial (minimum load level), a step-wise increase in load up to a relative maximum load level, and a return back to the initial value. As shown on the figure, each load cycle is repeated at least once using of the same load steps and load levels. The level of load achieved through each step and each cycle may be subject to change depending on the behavior of the structure under any given load. The number of cycles and the number of steps listed below should be considered as a minimum.

- *Benchmark*: At the beginning of the rapid load test, an initial reading of the instrumentation should be taken at least 1 minute after all the equipment is functioning and no load is being applied to the test member, other than the testing equipment. The benchmark is shown in Figure 9.26 as the constant line beginning at time zero and indicating no load. After initial values have been recorded, the load test commences with the first cycle.

- *Cycle A*: The first load cycle consists of load steps, each increased by no more than 10% of the maximum total load expected in the rapid load test. The load is increased in steps, until the service level of the member is reached, but no more than 50% of the maximum anticipated test load, as shown in Figure 9.26. At the end of each load step, the load should be maintained until the parameters, which define the response of the structure (e.g., deflections, rotations, and strains), have stabilized, but not less than 1 minute. The maximum load level for the cycle should also be maintained until the structural response parameters have stabilized, but no less than 2 minutes. Holding the test load until the structural response has stabilized shows that the member has "realized the effects of that load to an acceptable degree" (FitzSimons and Longinow, 1975). For response parameters to be considered stable, the recorded values in the second half of the time interval under a constant load should not exceed 15% of that attained in the first half of the time interval. If this condition is not met after the minimum time interval, the load should be maintained until such time as it is met or the member has been deemed unsafe. Changes in the applied load should be made "quasi-statically." Essentially, changes in load must be slow enough so that structural response parameters can be monitored. While unloading, the load should be held constant at the same load levels as the loading steps for at least 1 minute, as shown in the figure. In certain instances, it may be impractical to completely unload the member at the end of each cycle. In those cases, the load may be held at no more than 10% of the maximum anticipated test load, which is shown by the shaded region in Figure 9.26.
- *Cycle B*: The second load cycle, cycle B, is a replica of the first cycle, as shown in Figure 9.26. By duplicating a load cycle, one is able to check the repeatability of the structural response parameters at each load step. If a significant difference appears between two consecutive and equal cycles, the load test should be halted and the structure should be re-evaluated. The load cycles are also repeated, because the first load cycle is often used for the "bedding-in" of the system (Bungey, 1989).
- *Cycles C and D*: These are two identical cycles which achieve a maximum load level that is approximately half way between the maximum load level achieved in cycles A and B and the total anticipated test load. The load steps at the beginning of cycles C and D may be greater than 10%, but only up to the load level that was attained in cycles A and B. It is often helpful to repeat some of the load steps taken in the cycles A and B to again check the repeatability of the test member, as shown in Figure 9.26. As the level of load exceeds that attained in the cycles A and B, the load steps should again not be greater than 10% of the maximum expected test load. Each load level in cycles C and D is maintained until

the structural response parameters have become stable, but not less than 1 minute, and 2 minutes for the relative maximum.

- *Cycles E and F*: Similarly, the fifth and sixth load cycles should be identical, and they should reach the maximum anticipated test load, as shown in Figure 9.26. The load steps up to the load level attained in cycles C and D may be larger than 10%, but repeating some of the previous load levels allows for the repeatability check. Once the level of load becomes greater than that achieved in the cycles C and D, the load steps should not be larger than 10% of the maximum anticipated test load. Each load step should be repeated as the load is being decreased in each cycle.

- *Additional cycle*: In some instances, additional load cycles may be required in order to show the member's ability to hold the maximum test load. Additional cycles should be identical to cycles E and F as defined above.

- *Final cycle*: At the conclusion of the final cycle, the test load should be decreased to zero, as shown for cycle F in Figure 9.26. A final reading should be taken no sooner than two minutes after the entire test load – not including the equipment used to apply the load – has been removed.

### 9.3.2   *Analysis during testing*

Monitoring the structural response during testing to ensure stability of the system at every load level is key to the performance of a successful rapid load test. The stabilization of the structural response parameters under a constant load shows the member's ability to safely maintain that load. Monitoring additional parameters (i.e., repeatability, deviation from linearity, and permanency) during a rapid load test also gives an indication of the behavior of the test member.

- *Repeatability*: One method of rating a structure's performance during a rapid load test is by checking the repeatability of some structural responses, typically deflection. Repeatability, calculated using Equation (9.2) (with reference to Figure 9.27), is the ratio of the difference between the maximum and residual deflections recorded during the second of two identical load cycles to that of the first.

    During a rapid load test, if the load is not decreased to zero at the end of each cycle, the origin of the load versus deflection curve is shifted to $P_{min}$, as shown in Figure 9.27 to calculate repeatability.

$$\text{Repeatability} = \frac{\Delta^B_{max} - \Delta^B_r}{\Delta^A_{max} - \Delta^A_r} \times 100\% \qquad (9.2)$$

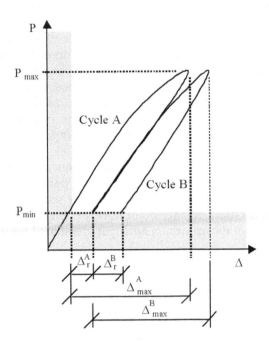

*Figure 9.27* Sample load–deflection curve for two cycles.

where:

$P_{max}$ = maximum load level achieved by cycles A and B
$P_{min}$ = minimum load level achieved at the end of cycles A and B
$\Delta^B_{max}$ = maximum deflection in cycle B under a load of $P_{max}$
$\Delta^B_r$ = residual deflection under cycle B under a load of $P_{min}$
$\Delta^A_{max}$ = maximum deflection in cycle A under a load of $P_{max}$
$\Delta^A_r$ = residual deflection under cycle A under a load of $P_{min}$

If a test member incurs a greater net deflection under a particular load the second time as opposed to the first, this may be an indication that the member has been softened. Experience has shown that a repeatability of greater than 95% is satisfactory. By checking the repeatability of deflections, one is not only monitoring the structure's behavior, but also gaining assurance that the data collected during the rapid load test is consistent.

- *Deviation from linearity*: This is a measure of the nonlinear behavior of a member being tested. As a member becomes increasingly more damaged, its behavior may become more nonlinear. The rapid load test method of applying load in cycles provides the opportunity to calculate the deviation from linearity in a variety of ways. The following illustrates how deviation from linearity is calculated for the load–deflection envelope of the test member.

To calculate the deviation from linearity, linearity must first be defined. Linearity is the ratio of the slopes of two secant lines intersecting the load–deflection envelope. The load–deflection envelope is the curve constructed by connecting the points corresponding to only those loads that are greater than or equal to any previously applied loads, as shown in Figure 9.28. This figure is a plot of load (on the vertical axis) versus deflection (on the horizontal axis), for a member undergoing a load test consisting of six cycles, labeled A through F.

Given a point $i$ with coordinates $P_i$ and $\Delta_i$, the secant line, shown as a dashed line in Figure 9.28 drawn from the origin to the point $i$ on the load–deflection envelope, and $\alpha_i$ is its slope. The reference secant line is the one that joins the origin to a reference point having coordinates $P_{ref}$ and $\Delta_{ref}$, where $P_{ref}$ is 50% of the maximum anticipated test load for the

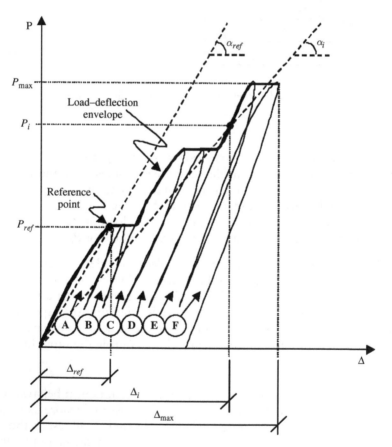

*Figure 9.28* Sample load–deflection curve for six cycles.

rapid load test, $P_{max}$. The linearity of any point $i$ on the load–deflection envelope is the % ratio of the slope of that point's secant line, $\alpha_i$, to the slope of the reference secant line, $\alpha_{ref}$, as shown in the following equation:

$$\text{Linearity}_i = \frac{\alpha_i}{\alpha_{ref}} \times 100\% \tag{9.3}$$

The deviation from linearity of any point on the load–deflection envelope is the complement of the linearity of that point, as shown in Equation (9.4).

$$\text{Deviation from linearity}_i = 100\% - \text{linearity}_i \tag{9.4}$$

Once the level of load corresponding to the reference load has been achieved, deviation from linearity should be monitored until the conclusion of the rapid load test. Experience (Mettemeyer, 1999) has shown that the values of deviation from linearity, as defined above, are less than 25%. Research is being conducted in this area to determine whether deviation from linearity can be used to predict failure. Deviation from linearity may not be useful when testing a member that is expected to behave in a nonlinear, but elastic manner. For such members, repeatability, as defined earlier, and permanency, as defined next, may be better indicators of damage in a tested structure.

• *Permanency*: The amount of permanent change displayed by any structural response parameter during any given load cycle is defined as permanency. Deflection permanency, calculated using Equation (9.5), may only be legitimately calculated for the second cycle of two identical load cycles, for example, cycle B in Figure 9.27. Often a system will seem to have suffered a much larger permanent deformation in the first of two identical load cycles because of the "bedding-in" of the system. In Equation (9.5), $\Delta_r$ is the residual deflection and $\Delta_{max}$ is the maximum deflection that has occurred in the member during a single cycle.

$$\textit{Permanency} = \frac{\Delta_r}{\Delta_{max}} \times 100\% \tag{9.5}$$

Experience has shown that a deflection permanency of less than 10% is acceptable. Bares and FitzSimons (1975) suggested that 25% permanency is acceptable for RC structures and 20% permanency for PC structures.

### 9.3.3 Interpreting the results

In some instances, it may be sufficient to show the member's ability to hold a given load as proof of successful performance. However, it is often helpful to perform a more in-depth analysis of the data.

### 9.3.3.1  Analytical modeling

Once the rapid load test has been conducted, analytical models can be used to verify assumptions regarding load sharing characteristics of adjoining members, effects of a concentrated load versus a distributed load, and degree of support fixity. The analytical model can be refined by adjusting the boundary conditions and stiffness of the adjacent elements to match the actual, measured, deformed shape of the structure. Once these refinements have been made, the analytical model may also be used to determine the internal forces caused by the test loads. For example, when testing in flexure a two-way slab with symmetric boundary conditions, a concentrated load is applied at midspan of the test element that will produce the same moment per unit width at the critical section (i.e., midspan) as that caused by a uniformly distributed load. Equation (9.6) is used to determine the magnitude of the uniformly distributed load that is simulated by the concentrated test load.

$$w_{sim} = \frac{P_T}{C_1 \times C_2 \times l_1 \times l_2} \tag{9.6}$$

The variables shown in Equation (9.6) are defined as follows:

$w_{sim}$ = uniformly distributed load simulated by the test load
$P_T$ = magnitude of a concentrated test load
$C_1$ = multiplier to account for contribution of adjacent elements
$C_2$ = multiplier to account for distributed load versus point load
$l_1$ = span in the primary direction
$l_2$ = span in the perpendicular direction

The coefficient $C_1$ accounts for the contribution of the adjacent elements (load sharing) to the member's response to loading. The value of $C_1$ is essentially the width of the slab that is effective in resisting the applied load on the test member. Based on deflection values collected in either a preliminary test or under low levels of load, the value of $C_1$ can be more accurately defined. Following the same example discussed earlier, the value of $C_1$ for a unit strip of a two-way slab with symmetric boundary conditions tested in positive flexure can be calculated using:

$$C_1 = \frac{l_1 \times \sum_{i=1}^{n} \Delta_i}{2 \times \Delta_1 \times b \times n} \tag{9.7}$$

The values shown in Equation (9.7) are defined as follows:

$\Delta_i$ = deflection measured at point $i$
$\Delta_1$ = deflection measured under the load

$b$ = width of the unit strip

$n$ = number of equally spaced deflection readings orthogonal to the span of the test member

The coefficient $C_2$ accounts for the difference between applying a concentrated load and a uniformly distributed load to the test member. For example, consider the case of a load test on a simply supported beam where the concentrated load is applied at midspan to produce a maximum positive moment equal to that of the same beam subjected to a uniformly distributed load. In this case, $C_2 = 0.5$. Values for $C_2$ can be computed for every level of fixity from 0%, simply supported, to 100%, fixed. The level of fixity for a given member can be determined by computing $R$, the ratio of the net deflection at quarter-span to that at mid-span. The highest value of $R$ corresponds to a span with pinned ends (no rotational stiffness) and the lowest value indicates a fully fixed support condition. The values of $R$ and $C_2$ can be plotted on separate vertical axes versus the % fixity of the supports. Figure 9.29 shows a sample curve for the case of a span with symmetric boundary conditions loaded with a concentrated load that produces a moment at mid-span equivalent to that due to a uniformly distributed load. Similar curves can be generated for other arrangements in loading, simulated response conditions, and/or boundary conditions.

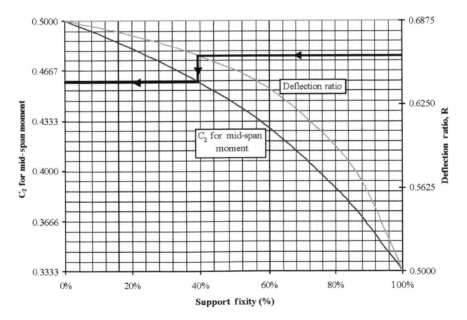

*Figure 9.29* Variation of $C_2$ and $R$ for a symmetrically bound member loaded at midspan.

Upon completion of a preliminary test, or lower load cycles, the value of $R$ can be computed as the average of all the values calculated based on measurements taken at load levels within the elastic range. Knowing $R$, fixity and $C_2$ can be found. By applying the proper level of support fixity to the analytical model, the shape of the analytical and experimental elastic curves will be similar. The match between measured and analytical deflection values can be obtained by adjusting the concrete elastic modulus assumed in the model. Once a satisfactory agreement between measured and analytical deflections is confirmed, the model is then used to accurately determine the magnitudes and variations of bending moments in the slab.

### 9.3.3.2  Analysis of data

Experience has shown that graphical as well as numerical efforts may aid in the understanding of the structure's response to a rapid load test. The most common graphical aids include curves displaying load and a structural response parameter plotted versus time and load plotted versus a structural response parameter. Numerical efforts include the calculation of permanency as a criterion for the acceptable performance of the test member.

- *Load and structural response parameters versus time*: A "time history" of a rapid load test is a graph of load and some structural response parameter plotted on dual vertical axes versus time. This graph shows the stabilization of that specific structural response parameter under a constant load. It can also be used to show the repeatability of the structural response of two identical load cycles.
- *Load versus structural response parameters*: A complete plot of the load versus any structural response parameter is a way to show the linear and elastic behavior of the test member. Members approaching failure may show signs of nonlinear or inelastic behavior. Residual deformations are also clearly shown. It is helpful to plot the theoretical response along with the actual results.
- *Permanency*: Permanency, as described above and shown in Equation (9.5), may be used as a criterion that defines the acceptable performance of a test member under a given load. Permanency as an acceptance criterion should only be that which is calculated for the final cycle of the rapid load test. Additional repetitions of the final cycle may be required. If the permanency value used as a criterion for acceptance is near the preestablished limit, one may wish to include the calculations of repeatability and deviation from linearity in the posttesting analysis.

### 9.3.3.3  Conclusions based on results

Once the numerical and graphical analysis has been completed, conclusions as to the acceptable performance of the member under the loads applied can be drawn. This method of testing only quantifies a safe level of load for a member under short-term conditions. Long-term behavior influenced by phenomena such as creep and degradation must be characterized separately from this testing procedure.

# References

ACI Committee 318 (1956) Building code requirements for reinforced concrete (ACI 318-56). *Journal of the American Concrete Institute*, 27(9), pp. 920–922.

ACI Committee 318 (1971) *Building Code Requirements for Reinforced Concrete (ACI 318-71)*. Farmington Hills, MI: American Concrete Institute, 78 pp.

ACI Committee 318 (2005) *Building Code Requirements for Structural Concrete and Commentary (ACI 318R-05)*. Farmington Hills, MI: American Concrete Institute, 430 pp.

ACI Committee 437 (1991) *Strength Evaluation of Existing Concrete Buildings (ACI 437R-91)*. Farmington Hills, MI: American Concrete Institute, 24 pp.

American Society of Civil Engineers (1991) *Guideline for Structural Condition Assessment of Existing Buildings (ASCE 11-90)*. Reston, VA: American Society of Civil Engineers, 89 pp.

Barboni, M., Benedetti, A., and Nanni, A. (1997) Carbon FRP strengthening of doubly curved, precast PC shell. *ASCE Journal of Composites for Construction*, 1(4), pp. 168–174.

Bares, R. and FitzSimons, N. (1975) Load tests of building structures. *ASCE Journal of the Structural Division*, 101(ST5), pp. 1111–1123.

Benedetti, A. and Nanni, A. (1998) On carbon fiber strengthening of heat-damaged prestressed concrete elements. In: *Proceedings, European Conference on Composite Materials (ECCM-8)*, I. Crivelli-Visconti (ed.), University of Naples, Naples, Italy, June 1998, pp. 67–74.

Bick, R.R. (1998) *Ocean Vista Power Generation Station Turbine Deck Rehabilitation Project*. International Concrete Repair Institute Award for Outstanding Concrete Repair Recipient, Southern California Edison, 26 pp.

Building Officials and Code Administrators International (1987) *The BOCA National Building Code*. 10th edn. Country Club Hills, IL: Building Officials and Code Administrators International, Inc., pp. 231.

Bungey, J.H. (1989) *The Testing of Concrete in Structures*. 2nd edn. New York, NY: Chapman & Hall, 228 pp.

California Department of Transportation (Caltrans) (2008a) *Feather River Bridge: Inclinometer*. [Online images] Available at: <http://www.dot.ca.gov/hq/esc/ttsb/instrumentation/photo_album.html> [Accessed 15 August 2008].

California Department of Transportation (Caltrans) (2008b) *Lighting Standards: Installing Gage*. [Online images] Available at: <http://www.dot.ca.gov/hq/esc/ttsb/instrumentation/photo_album.html> [Accessed 15 August 2008].

Carr, J.J. (1993) *Sensors and Circuits: Sensors, Transducers, and Supporting Circuits for Electronic Instrumentation, Measurement, and Control*. Englewood Cliffs, NJ: Prentice-Hall, Inc., 324 pp.

FitzSimons, N. and Longinow, A. (1975) Guidance for load tests of buildings. *ASCE Journal of the Structural Division*, 101(ST7), pp. 1367–1380.

Fling, R.S., McCrate, T.E., and Doncaster, C.W. (1989) Load test compared to earlier structure failure. *Concrete International*, 18(11), pp. 22–27.

Fraden, J. (1993) *AIP Handbook of Modern Sensors: Physics, Designs and Applications*. New York, NY: American Institute of Physics, 552 pp.

Genel, M. (1955a) Ripartizione laterale dei carichi in seguito alla monoliticita del cemento armato – Parte prima, *Il Cemento*, Associazione Italiana Cemento Armato, Vol. 6, pp. 6–15.

Genel, M. (1955b) Ripartizione laterale dei carichi in seguito alla monoliticita del cemento armato – Parte seconda, *Il Cemento*, Associazione Italiana Cemento Armato, Vol. 7, pp. 5–13.

Gold, W.J. and Nanni, A. (1998) In-situ load testing to evaluate new repair techniques. In: *Proceedings, NIST Workshop on Standards Development for the Use of Fiber Reinforced Polymers for the Rehabilitation of Concrete and Masonry Structures*, D. Duthinh (ed.), Tucson, AZ, January 1998, pp. 102–112.

Gold, W.J., Blaszak, G.J., Mettemeyer, M., Nanni, A., and Wuerthele, M.D. (2000) Strengthening dapped ends of precast double tees with externally bonded FRP reinforcement. In: *Proceedings, ASCE Structures Congress 2000*, Philadelphia, PA, M. Elgaaly(ed.), May 8–10, 2000, Philadelphia, PA [CD version, #40492-045-003].

Hall, W.B. and Tsai, M. (1989) Load testing, structural reliability, and test evaluation, *Structural Safety*, Vol. 6, pp. 285–302.

Hogue, T., Oldaker, L., and Cornforth, R.C. (1999a) FRP Specifications for the Oklahoma City Myriad. In: *Proceedings, Fifth ASCE Materials Engineering Congress*, L.C. Bank (ed.), Cincinnati, OH, May 1999. Reston, VA: ASCE, pp. 284–291.

Hogue, T., Conforth, R.C., and Nanni, A. (1999b) Myriad Convention Center floor system: issues and needs. In: *Proceedings, Fourth International Symposium on FRP for Reinforcement of Concrete Structures (FRPRCS4)*, Baltimore, MD, November 1999. C.W. Dolan, S. Rizkalla, and A. Nanni (eds), Farmington Hills, MI: American Concrete Institute SP-188, pp. 1145–1161.

Mettemeyer, M. (1999) *In-situ Rapid Load Testing of Concrete Structures*. Master's Thesis, Rolla, Missouri: University of Missouri – Rolla.

Mettemeyer, M., Serra, P., Wuerthele, M., Schuster, G., and Nanni, A. (1999) Shear load testing of CFRP strengthened double-tee beams in precast parking garage. In: *Proceedings, Fourth International Symposium on FRP for Reinforcement of Concrete Structures (FRPRCS4), SP-188*, Baltimore, MD, November 1999, C.W. Dolan, S. Rizkalla, and A. Nanni (eds.), Farmington Hills, MI: American Concrete Institute, pp. 1063–1072.

Nanni, A. and Gold, W.J. (1998a) Evaluating CFRP strengthening systems in situ. *Concrete Repair Bulletin*, 11(1), pp. 12–14.

Nanni, A. and Gold, W.J. (1998b) Strength assessment of external FRP reinforcement. *Concrete International*, 20(6), pp. 39–42.

Nanni, A., Gold, W.J., Thomas, J., and Vossoughi, H. (1998) FRP Strengthening and on-site evaluation of a PC slab. In: *Proceedings, Second International Conference on Composites in Infrastructure (ICCI-98)*, H. Saadatmanesh and M.R. Ehsani (eds.), Tucson, AZ, January 1998, Vol. I, pp. 202–212.

RDP Electrosense (2008) *DCTH Series DC to DC LVDT Displacement Transducer*. [Online images] Available at: <http://www.rdpelectrosense.com/displacement/lvdt/general/dcth-configuration.htm> [Accessed 15 August 2008].

RILEM Technical Committee 20-TBS. (1984) General recommendation for static loading test of load-bearing concrete structures in situ (TBS2). In: *RILEM Technical Recommendations for the Testing and Use of Construction Materials*, London: E & FN Spon, pp. 379–385.

Sawyer, T. (1998) Airport beam cracks are filled, wrapped in time for holidays. *Engineering News-Record*, McGraw-Hill, 21 December 1998, p. 16.

Wuerthele, M.D. (1999) CFRP strengthening at Pittsburgh International Airport's short-term parking garage. In: *Proceedings, Fifth ASCE Materials Engineering Congress*, L.C. Bank (ed.), Cincinnati, OH, May 10–12, 1999. Reston, VA: ASCE, pp. 276–283.

# Index